FROM ICE AGE
to
WETLANDS

First published in 2017 by Redshank Books

ISBN 978 0 9954834 0 8

Cover and Design by Helen Taylor

Printed in the UK by Short Run Press

Redshank Books
Brunel House
Volunteer Way
Faringdon
Oxfordshire
SN7 7YR

Tel: +44 (0)845 873 3837

www.libripublishing.co.uk

FROM ICE AGE
to
WETLANDS

The Lea Valley's Return to Nature

Jim Lewis

Jim Lewis

REDSHANK
BOOKS

Foreword by Terry Farrell

Jim Lewis is an extraordinary man. He is a traveller in that long tradition of indefatigable British explorers, journeying intellectually into unknown yet fascinating territory. Out of the glorious and chaotic metropolis which is London, Jim has discovered in his travels and revealed through his writing one of the great wonders of London – the extraordinary history of the Lea Valley.

The Lea Valley is the place where Jim spent his working life. The places he worked, and the characters he encountered there, drew him into the fascinating history of the place and inspired him to reveal the full story. I first encountered Jim through my own research into the Lea Valley, as part of my work in place making and characterisation of the Thames Gateway. Little did I know I could spend a lifetime struggling to learn only a fraction of what Jim has discovered.

Jim is a relentless advocate for this extraordinary place. For many years, he has been campaigning to seek recognition for the significance of the Lea Valley, as part of the rich history of London. For this small part of London changed the world – a crucible of scientific discovery and industrial firsts. His earlier books tell the unique story of the region, its scientists, engineers and entrepreneurs. But most significantly, Jim has revealed how the Lea Valley was the birthplace of the post-industrial revolution – the electronic technological revolution – which arguably began in the Lea Valley with the invention of the diode valve by Professor Ambrose Fleming. This small but inspired device allowed, for the first time, the control of a stream of electrons by electronic means, paving the way for modern electronic communication around the world and across the vast expanse of space.

Given the focus on the Lea Valley with the creation of the Queen Elizabeth Olympic Park, Jim is unique in the way he has recognised the significance of the place. In this new book, Jim discusses how the Lea Valley was formed and skilfully uses the region as an example to warn us of the damage that our consumer-driven lifestyles have created for our planet. Therefore, I urge you to journey with him through a past which is shaping the future and in so doing discover ways to protect our precious environment. It is people like Dr Jim Lewis who keep alive the magic of the place for present and future generations.

Sir Terry Farrell

ABOUT THE AUTHOR

Dr Jim Lewis has spent most of his career in the consumer electronics industry, apart from a three-year spell in the Royal Air Force servicing airborne and ground wireless communications equipment. When working in the Lea Valley for Thorn EMI Ferguson, he represented the company abroad on several occasions and was involved in the exchange of manufacturing technology. Currently, he is a consultant to Terry Farrell & Partners on the historical development of London's Lea Valley and a volunteer with social services teaching students who have learning difficulties. A freelance writer (with twelve published books), researcher and broadcaster for his specialist subject – London's Lea Valley – he also has a genuine passion for encouraging partnership projects within the local community, which in the long term are planned to help stimulate social and economic regeneration.

In 2012, Dr Lewis was appointed Contributory International Professor by the Clark H. Byrum School of Business, Marian University, Indianapolis for his work with students on the modern British service economy.

Dr Lewis is married with four grown-up children and lives in Lincolnshire

AUTHOR'S NOTE

In June 2016 I was invited by the London Wildlife Trust to lead a group on an industrial history walking tour. The tour was of a Thames Water reservoir site that was in the process of being developed as a new wildlife sanctuary to be called the "Walthamstow Wetlands". When completed, the site, located close to the Ferryboat Inn, Ferry Lane, Walthamstow, will be the largest wetland nature reserve in London.

At the time of the visit, work was well underway to remove the years of silt build-up from two of the low-level Victorian reservoirs that had been selected to form new wildlife habitats. The removed sediment was deposited towards the reservoirs' edge to create new reed beds. Once established, these beds will naturally improve water quality and provide wildlife habitats for birds, mammals and insects. It is also hoped that the

The author, centre, with BBC Essex, broadcasting from the Queen Elizabeth Olympic Park in 2012.

The author filming, on a wet and windy day, a piece for the Mayor of London's office. The film was distributed to international broadcasters attending the 2012 Olympics.

establishment of the reed beds will attract a greater variety of migrating and over-wintering birds to the site and in turn introduce these newcomers to other nature reserves within the region.

Witnessing this imaginative, community-focussed project taking place was my inspiration for this new Lea Valley book. Here it is planned to highlight the major events that have helped shape, not always in a positive way, the Lea Valley and its environs. Also, I will explain how scientists, engineers, developers and agriculturalists are coming together in their understanding of the importance of helping industry, agriculture and nature co-exist by developing new ways of protecting our diminishing natural resources.

Around the world, since at least the time of the Industrial Revolution, many people have accepted, or are beginning to accept, that it is our voracious appetite for more and more creature comforts that is seriously damaging our environment. Therefore, it is hoped that all nations will now take seriously such issues as climate change, melting ice sheets, de-forestation and declining animal and plant species; and take the necessary steps required to reduce the release of major pollutants, such as carbon dioxide and other noxious gasses and materials, into the atmosphere, and introduce imaginative schemes that protect our oceans, costal communities and wildlife. By taking such measures it should be possible to slow or halt the rise in global temperatures that scientists have shown causes the melting of polar ice which in turn increases sea levels, changes weather patterns and, amongst other things, causes flooding. Projects like the Walthamstow Wetlands should be seen as part of a global strategy that not only helps us to learn about our natural environment but also gives us the opportunity to protect our planet for future generations by encouraging politicians and industrialists to think conservation before planning new projects.

Jim Lewis

ACKNOWLEDGEMENTS

The author wishes to thank the following organisations, companies, societies and individuals for their encouragement, support and advice and for supplying many of the illustrations within this book:

Alexandra Palace and Park Trust, Wood Green, London

Bruce Castle Museum, Tottenham, London

Cath Patrick

Edmonton Hundred Historical Society, Enfield, Middlesex

Enfield Archaeological Society, Enfield, Middlesex

Enfield Local History Unit, Thomas Hardy House, London Road, Enfield, Middlesex

Epping Forest Museum, Sun Street, Waltham Abbey, Essex

Guy & Wright Ltd, Green Tye, Hertfordshire

Hertfordshire & Middlesex Wildlife Trust

Lea Valley Growers' Association, Cheshunt, Hertfordshire

Lee Valley Regional Park Authority, Myddelton House, Enfield, Middlesex

London Borough of Enfield, Civic Centre, Enfield, Middlesex

London Borough of Haringey, Civic Centre, Haringey, London

London Borough of Newham, Town Hall Annex, Barking Road, East Ham, London

London Borough of Waltham Forest, Town Hall, Forest Road, Walthamstow, London

Markfield Beam Engine & Museum, Markfield Park, Tottenham, London

Museum of London, London Wall, London

Professor Tony Travers, London School of Economics

RCHME Cambridge (National Monuments Record), Brooklands Avenue, Cambridge

River Lea Tidal Mill Trust, Bromley by Bow, London

Thames Water, Gainsborough House, Manor Farm Road, Reading, Berkshire

The Corporation of Trinity House, Tower Hill, London

The Greater London Record Office, Northampton Road, London

The Hackney Society, Eleanor Road, Hackney, London

The Herts & Middlesex Wildlife Trust, St Michael's Street, St Albans, Hertfordshire

The House of Lords Record Office, Westminster, London

The Institution of Engineering and Technology, Savoy Place, London

The Institution of Civil Engineers, George Street, London

The Institution of Mechanical Engineers, Birdcage Walk, London

The London Organising Committee of the Olympic & Paralympic Games, Canary Wharf, London

The London Wildlife Trust, Dean Bradley House, 52 Horseferry Road, London

The Natural History Museum, Kensington, London

The New River Action Group, Hornsey, London

The National Archive, Kew, Richmond, Surrey

The Pump House Steam & Transport Museum, South Access Road, Walthamstow, London

The Science Museum, Kensington, London

The Waltham Abbey Royal Gunpowder Mills Company Ltd, Waltham Abbey, Essex

Tomsworld Ltd, Pecks Hill, Nazeing, Essex

Tower Hamlets Local History Library, Bancroft Road, Tower Hamlets, London

Bird photographs: Brenda Chanter, Brian Anderson, Dennis Meadhurst, Graham Canny, Ken Bentley, Mark Braun, Paul Lister, Steven Swaby and Tim Hill

While many individuals have freely given their knowledge, some unknowingly, which has contributed greatly to the production of this book, I have, on a number of occasions paid special tribute to certain people in the footnotes of various chapters.

I could not let the occasion pass without recording my sincere thanks to my wife Jenny for her superb editorial skills and outstanding patience. The author freely admits that this voluntary sacrifice on Jenny's part has comprehensively tested the cement that holds our wonderful marriage together.

CONTENTS

FROM ICE AGE TO WETLANDS
the Lea Valley's Return to Nature

The area between north and north east London through Harlow to Cambridge has emerged as the UK's innovation corridor which has genuine potential to be a rival to the global centres of technology and life sciences in San Francisco and Boston Massachusetts.

The Lea Valley is at the heart of this corridor and that is no accident: the innovation which drove developments such as the UK's first motor car, first powered flight and the birthplace of the modern electronics industry continues to drive this corridor today. But Britain's success in the new knowledge economy will only happen if it is placed in the right environment where people with talent, ideas and ambition want to work, live and play.

The Lea Valley is that place and this book shows how it has evolved over thousands of years and remains important to the future of London and the rest of the country.

John McGill
Director LSCC
London, Stansted, Cambridge, Corridor – sponsors of this book

INTRODUCTION

Those readers who know me and have journeyed with me over the years through the wealth of Lea Valley industrial history stories will no doubt appreciate that I am about to enter areas of the region's history that are outside my industrial expertise. Nevertheless, I believe the challenge should not be avoided as the quest for new knowledge will hopefully provide a platform to launch others down the research path and encourage them to uncover more about our region's best kept secrets.

To be able to understand the various influences that have shaped the Lea Valley it will be necessary to revisit some of the stories that I have already written to place the region within the timeline of historic events. Only by introducing the industrial history of some of the region's prominent sites will we be able to understand more clearly the transition from ignorance to the embracing and preservation of nature. In fact, when I started to construct the narrative for this book my earlier researches fell into place beautifully, allowing me to appreciate how much the various organisations and businesses of this relatively small region were, and still are, making major contributions towards the improvement of the environment.

THE POLITICAL DEVELOPMENT OF LONDON'S LEA VALLEY – AN OVERVIEW BY PROFESSOR TONY TRAVERS

Historical Background

Today's Lea Valley is the product of its social, cultural and economic origins. The Lee Valley Regional Park Authority (LVRPA), now 50 years old, inherited a river, its people and their collective history. The power of rivers and the economic impact of valleys is the subject of much academic and popular interest. Before embarking on the role the Authority has played in the recent development of the lower Lea Valley, it is worth considering the longer-term history of the area.

The Lea's sister, the Thames, is one of the world's best-known rivers. The Thames's economic and cultural power has contributed to an under-appreciation of its several tributaries. Much the same could be said of, say, the Passaic and Hackensack rivers in New York and New Jersey when compared to the mighty Hudson. The junction of the Lea (or Lee) and the Thames, appropriately at Leamouth, is at a point not only where two stretches of water merge, but also where two stories meet.

The Lea starts life at Leagrave, a leafy suburb of Luton. There is a modest plaque on posts in a small railed enclosure to mark the spot. Luton derives its name from that of the river added to the Saxon word for settlement, a 'tun'. 'Lea' may have been a Celtic word for 'bright river'.[1] It therefore rises in the Chilterns, which suggests a rather different image from the one it has gained from passing through its industrial heartlands to the south. It flows onwards through Harpenden and Welwyn Garden City, before meandering eastwards to Hertford. Here it divides into two, with a canalised river running in parallel to the ancient one for much of its further journey to London. The canalised section takes the name 'Lee', while the original river is still called the 'Lea'. Continuing south, it passes through Hoddesdon, Broxbourne, Cheshunt and Waltham Cross. It then dips under the M25 and on across the London boundary into Enfield, Waltham Forest and then down to the Thames. At 58 miles long, it is three-quarters the length of the Tyne, or four-fifths of the Mersey. It is thus one of the country's longer rivers.

The Lea's history created an economic and physical heritage which generated the impetus for the LVRPA. In Anglo-Saxon England, the Lea ran through the Kingdom of Essex. The river eventually became the border between Essex and Mercia, which defeated the former and annexed part of its territory.[2] The river was subsequently to mark the border between the counties of Hertfordshire and Essex, and also between the County of London and Essex.

In common with most of England, the Lea Valley area was significantly agricultural for the centuries before the Industrial Revolution. There was early industry in Luton, while Waltham Cross was both a market town and also a centre for the manufacture of gunpowder.[3] In the late seventeenth century, the Lea Valley was home to a number of pretty villages within what is today London: "Hackney long remained a village, its single church large enough to take the whole congregation… Poplar, Old Ford, Clapton, Stratford and Leyton – all these also remained picturesque villages. Bow stretched over the Lea

1 Lambert, Tim, *A Brief History of Luton*, http://www.localhistories.org/luton.html (Accessed 20th January 2017)

2 Yorke, Barbara, *Kings and Kingdoms of Early Anglo-Saxon England*, London: Routledge, 2002, pp.47–52

3 Cocroft, Wayne D., *Dangerous Energy: The Archaeology of Gunpowder and Military Explosives Manufacture*, Swindon: English Heritage, 2000

bridge to Stratford, towards Stratford and Bromley's high street hugged the River Lea. To the south, the Isle of Dogs remained the rough pasture it had been since the middle ages".[4] Leyton, like Luton, owes its name to the river added to the word for a town or settlement.

But the Lea Valley's position north of London and close to ancient roads to the North contributed to the siting of new industries and a transport corridor on this side of the growing capital. It was the rapid industrialisation of this corridor during the Industrial Revolution and the parallel expansion of London that laid the ground for particular needs of the Lea Valley from the middle of the twentieth century onwards.

As the country sought to think forward to the reconstruction and development that would be necessary after the Second World War, the government commissioned a review of the future planning requirements of the London area. Sir Patrick Abercrombie's *Greater London Plan 1944* was published in 1945, outlining an approach to the management and development of 'Greater London' and its surrounds. During the period from 1918 to 1939, London had sprawled outwards to cover some 700 square miles of south-east England.

The *Greater London Plan 1944* summarised the industrial history of the Lea valley thus:

> With a marked outward move in population up the Lee Valley from the congested East-End Boroughs taking place as far back as 1880, industry soon followed, stimulated by the skilled labour pool thus created there and by good rail and water transport facilities. The industries of this period were predominantly those of the East End group, i.e., furniture, clothing and metal working; subsequent to 1918, expansion, mostly in medium sized factories, has been rapid and through industry becoming more diversified with a big increase in such types as electrical engineering and the production of consumer goods; the area is not engaged in the production of retail luxury and semi-luxury articles to anything like the same extent as many industries in the western sector of London.[5]

The report explained at some length why so many nurseries were located in the area. Soil quality, water availability and the proximity to London markets such as those at Covent Garden and Spitalfields were seen as key advantages. Birmingham and Manchester were also supplied. Grapes, among many other crops, were grown. The area around Waltham Abbey, Cheshunt and Hoddesdon was seen as needing protection from the sprawl emanating northwards from Tottenham and Edmonton.[6]

The Plan envisaged a ban on industrial expansion within Greater London. "The ban on new industry in the Lea Valley will not only prevent over-industrialisation but will enable what is left of the low-lying valley lands to be welded into a magnificent open-space system, providing a more or less continuous green wedge between the densely populated East-End areas and the open country to the north.[7]

Finally, the *Greater London Plan* stated: "The Lee Valley gives the opportunity for a great piece of constructive, preservative and regenerative planning". While dismissive of the section from Stratford to the Thames as being "beyond salvage, cluttered up with heavy industry, gasworks and a network of railway

4 Porter, Roy, *London: A Social History*, London: Penguin, 2000, p.143

5 Abercrombie, Sir Patrick, *Greater London Plan 1944*, London: HMSO, 1945, pp.41–2

6 Ibid., p.42

7 Ibid., p.56

sidings", the report was enthusiastic about the stretch north of Stratford. Parks, open space, marshes, forests, sports fields, allotments, greyhound tracks, reservoirs, nature, footpaths and fishing were among the joys of the Lea's northern 20 or so miles. "These open areas are a great recreational and open-air lung to the crowded East End; their preservation is essential".[8]

Over 20 years later, the *Greater London Development Plan*, produced by the new Greater London Council in the late-1960s, stated:

> Within Greater London, the physical constraints and nature of the [newly created Lee Valley Regional] Park necessitate [an] approach with the emphasis on the provision of specific facilities rather than on the natural resources of the valley… a series of centres to be developed intensively for recreation, located near access points to the main road, bus and rail routes. A number of areas are scheduled for early development, at Pickett's Lock (Enfield), a riding playground at Three Mills, and the first school in the Hackney area, an adventure stage of a marina project at Springfields (Hackney). Some of the centres will be specialised, some multi-purpose, and will provide a wide range of indoor and outdoor recreational, cultural and entertainment activities.[9]

Both the *Greater London Plan* and the later *Greater London Development Plan* saw merit in protecting and preserving the Lea Valley. Both saw a difference between the more rural, northern, stretch of the river and the industrialised south. But by the late 1960s, the GLC was outlining a particular future for the London section which, by this time, was in the hands of the new Lee Valley Regional Park Authority.

Infrastructure and the Governance of Water

The Lea Valley, like many river valleys, had been chosen as a location for major infrastructure. The A10 runs through it on its way from London to Cambridge and Norfolk. The Great Eastern Railway and its predecessors built a line from Bishopsgate in the City of London up the Lea Valley towards Broxbourne, Cambridge and East Anglia. A link was built from Stratford to Broxbourne. Today, two lines run in parallel on the western side of the valley.

A system of reservoirs was built in the late nineteenth century following a Royal Commission on Metropolitan Water Supply. The New River Company drew water from the river at Hertford to supply north London. The western side of Essex relied on the Lea because the lower Thames was too close to the sea to be usable for fresh water. The Lea was a big enough river to supply the growing eastern side of London and western Essex, though like the Thames, it suffered from being both a sewer and a water source. There could also be shortages in dry periods. The East London Water Company struggled throughout its life to cope with both the shortages and with water quality by building small reservoirs at Old Ford (in Tower Hamlets). These early reservoir facilities were eventually replaced by ones at Lea Bridge. In the 1850s, larger reservoirs and filter beds were constructed at Walthamstow. But as the population and industrial base of east London and urban Essex expanded, the Lea became incapable of maintaining continuous water supplies. At this time, the famous Truman Brewery in Brick Lane used water from the Lea.

The Metropolitan Board of Works (MBW) and the London County Council (LCC) made a number of unsuccessful efforts between the 1870s and the 1890s to gain legislative powers to take private water companies into public control. Local authorities outside London resisted any suggestion that their

8 Ibid., p.105

9 *Greater London Development Plan Report of Studies*, London: Greater London Council, undated (probably 1968), pp.304–5

water should be supplied by the MBW or LCC (i.e. by a London-only authority). As a result, chaotic and weakly regulated water supply continued.

In 1901, the East London Co. obtained powers to build two massive reservoirs at Chingford. A year later, the Metropolis Water Act was passed, creating a new utility to deliver water services. The East London Co. and a number of other enterprises supplying water to the London region were bought out in 1903–4 and their undertakings transferred to the new Metropolitan Water Board (MWB), which was, effectively, a joint committee of local councils in London, Essex, Kent, Surrey, Hertfordshire and Middlesex. One member was appointed by the Conservators of the Thames and one by the Lee Conservancy Board. Ware, on the Lea, provided the northern limit of the MWB's territory.

Characteristically for Victorian governance improvements in and around London, the creation of the MWB took decades and ended up with a sprawling joint committee. Having said this, the Lea, the Thames and other rivers were now (at least in terms of water supply quality) within the control of a single, relatively powerful, authority. Earlier efforts at providing governance to the Lea had taken a number of forms.

Administering the River Lea

As early as 1424, the navigation of the river was in the hands of the Trustees of the River Lea, who were incorporated by a series of Acts of Parliament. The trustees were empowered "for the purpose of clearing, scouring and amending the river Ley".[10] The Lea was navigable by 1570. Legislation in that year empowered "the Lord Mayor Commonality and Citizens of London and their successors… to bring the said river of Lee from the town of Ware to the north side of the city of London".[11] The City of London thus assumed responsibility for the management and exploitation of the river. An Act of 1739 increased the number of trustees, giving the Lord Chancellor the power to appoint four members from the City of London and four each for the counties of Middlesex, Essex and Hertfordshire. Navigation along the river was open to all, but subject to the payment of fees and duties. The Trustees also received fees for supplying water to the New River Company and the East London Water Company.

In 1868, the Lee Conservancy Board (LCB) was created to replace the Lee trustees in relation to the Lee Navigation. Members were appointed by the outgoing Trustees, the Metropolitan Board of Works (MBW), private water companies, the Board of Trade, the City of London and the Stort Navigation trustees. The MBW was subsequently replaced by the London County Council and the water companies by the Metropolitan Water Board. The LCB took over responsibility for the canalised parts of the Lea and, from 1911, the River Stort Navigation. (The Stort joins the Lea near Hoddesdon, creating a navigable link from Bishop's Stortford to the Thames at Leamouth.) The LCB was responsible for reducing pollution and protecting water quality.

Legislation in 1930 passed part of the LCB's duties to the Lee Conservancy Catchment Board (LCCB) which became responsible for water supply, fisheries, pollution and drainage. The LCCB consisted of all LCB members plus six additional ones. The LCCB sent one member to the Metropolitan Water Board. In 1947, the LCB's responsibilities were transferred to the Lea District of the newly created British Transport Commission. The Commission was abolished in 1962 and its responsibilities distributed among four new organisations, one of which was the British Waterways Board (BW). The latter created a regional structure. BW was eventually abolished in 2012 and waterways transferred to the Canal and River Trust, a charity.

10   'History of Managing the Lea', the view from the bridge, Lea Bridge heritage, http://www.leabridge.org.uk/rivers-bridges-and-weirs/river-lee-trustees-and-lee.html (Accessed 12th February 2017)

11   Ibid.

In 1975, the new Thames Water Authority (TWA) took over the duties of the LCCB, as well as all the Metropolitan Water Board's responsibilities. When the TWA was privatised in 1989, the National Rivers Authority became responsible for the regulation of inland water, sewers, flood risk management, fishing, pollution and ecology. The NRA was then absorbed into the Environment Agency in 1996.

The evolution and role of the Lee Valley Regional Park Authority is the subject of this book. Suffice to say that the long and complex history of the ownership and management of the Lea, the Lee Navigation and its hinterland is revelatory. The river has long been of vital importance to London, Essex and Hertfordshire yet it has, over centuries, been randomly bundled into structural and administrative reorganisations with little thought as to its existence as a place with a purpose. In this sense, it is remarkably different from a body such as the City of London which has enjoyed consistent self-governing status for centuries.

The Lobby for a Lea Valley Authority

Several centuries of ad hoc river management arrangements for the Lea are outlined above. By the early part of the twentieth century, there was a small political lobby to manage and improve the Lea Valley. The most prominent of the early supporters was Herbert Morrison, who went on to be leader of the London County Council from 1934 to 1940, then subsequently a senior cabinet minister during Winston Churchill's wartime coalition and also during Clement Attlee's post-war Labour government. Morrison was elected a member of Hackney metropolitan borough council in 1919 and became mayor for 1920-1. In 1922, he was elected a member of the LCC and subsequently (in parallel) became an MP. No politician has even been as powerful in the capital. He was MP for Hackney South from 1923 to 1945, with a break from 1924 to 1929.[12]

Morrison's early political concerns included lobbying for reducing pollution and improving the use of the Lea Valley. As a Hackney councillor and MP, the Lea was a feature of the local landscape. Another Labour politician from Hackney, Sir Lou Sherman, was to prove instrumental in the genesis of the Lee Valley Regional Park Authority. This part of the story is picked up in the next chapter.

As outlined above, Sir Patrick Abercrombie's *Greater London Plan* explicitly proposed the use of the Lea for recreation, and also for banning new industry within the green parts of the river valley. It did not explicitly call for the creation of a separate authority. But the logic of protecting and improving the Lea and its environs can be dated to the Plan which, in turn, owes much to the prescience of Herbert Morrison and others who had proposed cleaning up the river. It is perhaps unsurprising that the 1960s, a decade which led to huge social, economic and political change in Britain, was the time when the formal machinery of a combined authority was created.

1966 and All That

The Lee Valley Regional Park Act received Royal Assent in December 1966. The Lee Valley Regional Park Authority, created by this legislation, emerged at a time of remarkable upheaval in Britain and the wider world. Harold Wilson's Labour government had won a majority of 96 in the general election in March of that year. In April, *Time* magazine ran its immortal 'Swinging London' cover story, which explained:

12 Donoughue, Bernard, and Jones, G.W., *Herbert Morrison: Portrait of a Politician*, Littlehampton Book Services Ltd, 1973

In this century, every decade has had its city. The fin de siècle belonged to the dreamlike round of Vienna, capital of the inbred Habsburgs and the waltz. In the changing '20s, Paris provided a moveable feast for Hemingway, Picasso, Fitzgerald and Joyce, while in the chaos after the Great Crash, Berlin briefly erupted with the savage iconoclasm of Brecht and the Bauhaus. During the shell-shocked 1940s, thrusting New York led the way, and in the uneasy 1950s it was the easy Rome of la dolce vita. Today, it is London...[13]

The public administration of London had recently been reformed. The London boroughs, which were elected for the first time (in shadow form) in May 1964, had fully assumed power on 1st April 1965. In June 1966, the counties and districts outside London were put under structural review by the Redcliffe-Maud commission.[14]

The Lea Valley legislation was not the only law to facilitate change to the fabric of London and the South East during 1966. The Covent Garden Market Act, 1966, allowed for the removal of the famous fruit market from its central London location to Nine Elms, Vauxhall. In 1965, the government had imposed a ban on new office building in central London. In 1967, the government decided to transfer the control of London transport to the newly formed Greater London Council.

Beyond local administration, other legislation passed during 1966 gave independence to Botswana, Lesotho and Barbados. The newly empowered Wilson government was to pass a series of socially liberal laws during 1967, fundamentally changing British society.

Thus, the LVRP came into existence at a time of reform and change within Britain. Its creation showed the government was prepared to innovate with administrative machinery so as to allow the improved government of a geographical area which paid little heed to the structure of district or county government. The River Lea, as outlined above, had provided an ancient boundary between Essex and Hertfordshire and also between the London County Council and Essex. The 1965 London government reforms retained the Lea as the boundary between (to the west) Enfield, Haringey, Hackney and Tower Hamlets, and (to the east) Waltham Forest and Newham. It remains the border between inner and outer London, as well as between the East and South East regions of England.

13 *Time*, 15th April 1966

14 Royal Commission on Local Government in England, 1966–9, Cmnd 4040, London: HMSO, 1969

TIME TRAVELLING THROUGH LONDON'S LEA VALLEY FROM PRE-HISTORY TO THE PRESENT

When giving talks on the industries of the Lea Valley, I am often asked to explain how the region came to be formed. Questions are also raised about those early people who came to the region to conquer and settle. To do these requests justice would require a very weighty tome indeed, but in the following few hundred words I shall try to describe how the valley came to be born.

Animal scull extracted from Lea Valley gravel workings.

Scientists have estimated that the earth is around 4,600,000,000 years old. Compared to this, the Lea Valley, which was formed 1,000,000 to 1,800,000 years ago, during the four glaciations (ice ages) of the Pleistocene Epoch, is a mere newcomer. The last ice sheet, which geologists have calculated advanced from the north to a line that extended west–east across Britain from south Wales through Finchley to south Essex, retreated about 11,000 to 10,000 years ago. Water created as the ice melted (melt water) brought down deposits of sand, gravel and clay that eventually formed the flat marshy flood plain of the lower Lea Valley. Evidence of this can be seen today in places like Walthamstow and Tottenham marshes. Indeed, we are fortunate that our recent ancestors named one of these areas Walthamstow Common Marsh and used it for growing hay and grazing cattle. This has allowed much of the Marsh's ancient character to remain. Since 1985, the area has been designated a Site of Special Scientific Interest (SSSI) and supports some 400 different species of insect, plant and animal.

Amwell gravel pits, where large extractions took place, is now a nature reserve.

Sand and gravel extraction on an industrial scale.

Sarcen stone, known as silcrete, removed from Amwell Quarry.

Artist's impression of carboniferous period.

St Albans gravel truck.

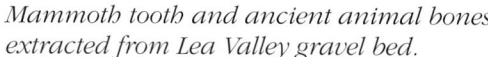

Mammoth tooth and ancient animal bones extracted from Lea Valley gravel bed.

Mammoth bone extracted from Lea Valley gravel bed.

Towards the northern end of the Lea Valley, which, unlike its southern counterpart, has not remained undisturbed, there is evidence of a much earlier period going back many millions of years. During the extensive gravel excavations that took place in the twentieth century, several types of fossil and a variety of prehistoric animal remains were unearthed, providing clues about the valley's early history for the archaeologists to study. The method used for extracting gravel is very much an industrial operation, which has meant that many of the ancient finds that have come to light failed to receive the same level of sensitive recording and care as would have been expected from a planned archaeological dig. However, on the plus side, since gravel extraction has now ceased in the upper Lea Valley, the aftermath has created a number of water features, which are home to a variety animals, insects and plants, not to mention the different species of fish.

Life-sized model of a woolly mammoth.

Imagine, if you will, the Lea Valley landscape some 10,000 years ago as the ice retreated northward. It would be possible to recognise the general layout of the high ground as being similar to that of today. However, the valley slopes and floor would have looked quite different. This earlier hostile landscape, devoid of trees and having a climate akin to that of Iceland, is a far cry from the present familiar topography of the Lee Valley Regional Park with its nature reserves, water features and other visitor attractions. Geological evidence suggests that the River Lea, at this time, was considerably wider than it is today, probably reaching a mile or more in places.

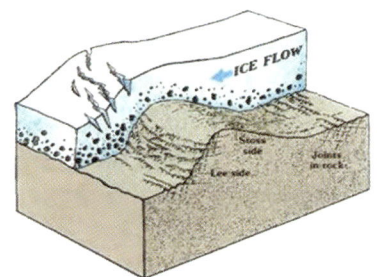

Ice-flow diagram.

As the region gradually became warmer and trees and vegetation began to flourish, we come upon evidence of habitation by man. Archaeologists excavating in the 1920s just east of Broxbourne discovered a Mesolithic (middle stone age) site (circa 6,000 to 3,000 BC). These early people, who appear to have been hunter-gathers, probably came to Britain over the land bridge which linked Britain to the Continent, until it was destroyed by natural causes, around 6,000 years ago. Indications of their existence (flint tools, animal remains and dwelling materials) were discovered buried beneath the Lea Valley peat. Further finds came to light in the 1970s when excavations at Broxbourne and Dobbs Weir revealed flint axes and a fish spear complete with flint barbs.

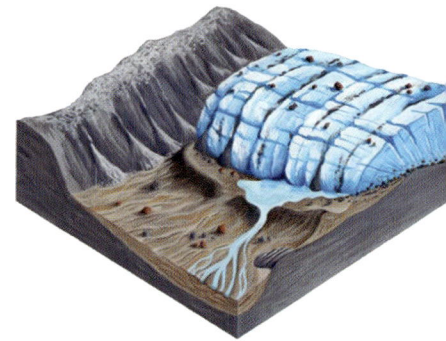

Artist's impression of the retreating ice sheet.

Map showing Ice-age coverage.

Land bridge with Continent.

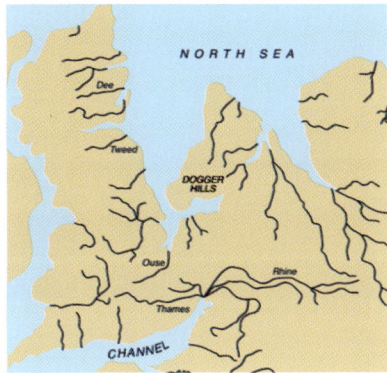

Doggerland map showing land bridge between Britain and Europe.

Verulamium Museum, St Albans, a good place to learn about the Roman occupation of Britain.

By around 3,000 BC the climate had warmed considerably and the Lea Valley landscape had changed from its former barren state to a more welcoming environment. Now the slopes were heavily wooded and the valley floor was covered by marshes on a bed of peat. The Neolithic (new stone age) people, who had crossed from the Continent to Britain, were beginning to enter the Lea Valley. These people had developed new skills that differed considerably from their Mesolithic forebears. They cleared the land, kept cattle, made pottery and planted cereals (wheat and barley). Essentially, they were farmers rather than hunter-gatherers.

The period 1,800–600 BC has been termed the Bronze Age. Here, for the first time, we see a giant leap in technology with the development of copper and bronze tools, drinking beakers and other implements. It is thought that this technology came to Britain through continental traders or possibly invaders. Finds by archaeologists, particularly in north Hertfordshire and more recently at Turnford (before the erection of the new housing estates), show that Bronze Age people were once in the region.

The beginning of the Iron Age came at around 600 BC and saw the first implements made from this new material. Iron tools were tougher and cheaper than copper or bronze. Now iron axes and ploughs tipped with iron allowed land clearance and cultivation to progress much faster than before. In the late 1960s, one of the earliest Iron Age settlements to be found in south-east England was discovered north of Ware.

When the Romans occupied Britain in 43 AD, the Lea Valley and what is now the Lee Valley Regional Park would have presented the invading army with a formidable obstacle to negotiate. To reach Camulodunum (Colchester), which we know was to become an extremely important Roman garrison town, the River Lea and its marshy extremities had to be crossed. Imagine the problems that this natural barrier posed for the Roman army, complete with equipment and stores as it marched north-eastward from the area that grew to become Londinium (London) in circa 50 AD. The Roman occupation of Britain appears to have lasted to around 450 AD and it is known, from the archaeological evidence (pottery, glass, coins, jewellery, tools, building materials, etc.) found within the Lea Valley and the surrounding countryside that the region must have held a position of considerable strategic importance. Ermine Street led north, from the Roman settlement of Londinium (London) directly along the valley floor to Ware, then continued northwards to Eboracum (York). Watling Street ran northwest to Verulamium (St Albans) then onward to Deva (Chester).

Scholars agree that the Roman occupation of Britain did not end abruptly, but gradually tailed off, allowing an overlap with the Saxon period that had begun around 400 AD. From genetic studies carried out by today's scientists, we are able to deduce that during this relatively long period of overlap, integration of the various peoples took place. This mixing of these different cultures would have also meant the sharing of knowledge, which has probably helped mould the British character into something that outsiders see as unique.

Archaeologists have yet to find sufficient Saxon remains, within the boundaries of the Lee Valley Regional Park, to construct a complete picture of the lifestyle of these early Germanic people in the area. However, in 1972 a late Saxon hoard of some seventeen artefacts was discovered in the gravels of Nazeing Marsh and in October 1987 a late Saxon log boat was unearthed from a site near the River Lea at Clapton, adjacent to Springfield Marina. It is to be hoped that the future will yield more finds that will further add to our knowledge.

By turning to the Anglo-Saxon Chronicle, written between the ninth and tenth centuries AD, we discover references to the next invaders, the Vikings (or Danes), who, by the ninth century, were active in the Lea Valley. In 886 AD, a bitter battle had taken place between the Saxons and Danes, which resulted in Alfred recapturing London. Afterwards, Alfred made an agreement with the Danish King Guthrum and a boundary was established between their respective territories. To achieve this, an imaginary line was draw along the Thames from the North Sea and then up the River Lea to near its source. According to the Chronicle, in the year 894 AD a large Viking fleet sailed up the River Thames and then up the River Lea to a point twenty miles north of London where a fortified camp was built with ditches and ramparts. It is recorded that King Alfred deprived the Danes of escape by blocking, or possibly lowering, the river. The Danes abandoned their camp, leaving their boats behind, and escaped by sending their women and children across country to East Anglia while the men marched overland to Bridgenorth in Shropshire.

Today's casual visitor to the Lee Valley Regional Park would probably find it difficult to imagine the events that took place in this relatively small area of land adjacent to the capital that helped shape a future British nation. It might also be hard to imagine, while taking a leisurely stroll in the now much altered and landscaped park, that the east bank of the River Lea was once Danish territory (Danelaw) while the west bank was Saxon (Wessex).

Restored Oseberg Viking Ship, Oslo Museum; if Vikings had come up the River Lea, it would probably have been in boats like this.

The influence of the Norman Conquest in 1066 on the Lea Valley region has left much material evidence for our visitors to see. Apart from the recording of several Lea Valley mills in the Domesday survey of 1086, there is also much remaining architecture to be observed. For example, the church at Waltham Abbey (the Abbey Church of Waltham Holy Cross and St Lawrence) although much altered over the centuries, has several fine Norman features. These include arches, a north and south door and a nave. Harold, Earl of Wessex, who later became the last Saxon King of England, founded the church circa 1060. After Harold's death at the battle of Hastings in 1066, it is said that his body was brought to Waltham Abbey for burial. There is currently a stone slab that stands to the east of the church that marks the grave. However, historians and archaeologists have suggested that the placing of the stone may be inaccurate, as Harold's body was reputedly moved on at least three different occasions. Perhaps modern archaeology should be brought in to

Waltham Abbey (the Abbey Church of Waltham Holy Cross and St Lawrence).

Map showing Danelaw and Wessex, where the River Lea formed the southern boundary.

Harold's grave at the east end of Waltham Abbey Church.

settle the argument? However if the results of a dig showed no evidence of Harold's remains, perhaps the Lea Valley would lose the claim to his resting place and visitor numbers could go down!

REFERENCES

Ashby, Margaret, *The Book of the River Lea*, Barracuda Books Ltd, 1991

Author unknown, *Walthamstow Marsh – a Guide to the History of the Area*, Lee Valley Regional Park Authority, 1986

Authors unknown, *Anglo-Saxon Chronicle*, Macmillan, 1982, translated by Anne Savage

Bascombe, K.N., and Bentley, John, *A Walk Round Waltham Abbey*, Waltham Abbey Historical Society, 1998

Batley, J., 'The Compilation of the Anglo-Saxon Chronicle Once More', *Leeds Studies in English* 16, 1985

Burnby, J.G.L., and Parker, M., *The Navigation of the River Lee (1190–1790)*, Edmonton Hundred Historical Society, Occasional Paper New Series No.36

Camp, John, and Dean, Dinah, *King Harold's Town*, Waltham Abbey Historical Society, 1988

Gibbs-Smith, Charles H., *The Bayeux Tapestry*, Phaidon

Heath, Cyril, *The Book of Amwell*, Barracuda Books Ltd, 1980

Higgins, Eric, and Bascombe, Kenneth, *Waltham Abbey*, Pitkin Pictorial Ltd, 1995

Higham, N.J., *The Norman Conquest*, Sutton Publishing Ltd, 1998

Humble, Richard, *The Saxon Kings*, Weidenfeld and Nicolson, 1980

Kiln, Robert, *The Dawn of History in East Herts*, The Hertfordshire Archaeological Trust, 1986

Morris, Carole A., 'A Late Saxon Hoard of Iron and Copper-Alloy Artefacts from Nazeing, Essex', *Medieval Archaeology*, Vol. XXVII, 1983

Saklatvala, Beram, *The Origins of the English People*, David & Charles, 1969

Thackray, John, *The Age of the Earth*, Her Majesty's Stationery Office, 1980

THE RIVER LEA – A WATERWAY THAT LINKS US TO THE PAST

At the start of its fifty-eight-mile (98 km) journey to the River Thames, the River Lea rises on Leagrave Common north of Luton, Bedfordshire, from several small springs that feebly force their way through the chalk of the Chiltern Hills. This small sample of the Lea's first water has been trapped in underground aquifers for thousands, perhaps millions of years. Depending on the season, the trickle slowly meanders in a shallow channel across marshy ground skirting Wauld's Bank, a four-thousand-year-old Neolithic henge, before being joined by its first tributary, Houghton Brook. More of a stream at this stage, the River Lea leaves its rural beginnings and crosses the Icknield Way, an ancient track that runs one-hundred and five miles (169 km) from the Ivinghoe Beacon, Buckinghamshire, in the west to Thetford, Norfolk in the east. The stream then makes its way beside the main road, passing lines of houses before entering the lakes in Wardown Park. On leaving the park, the water is taken below Luton in a culvert, exiting the other side of the town close to the Vauxhall motor factory. This factory came to Luton in March 1905, when it took over from Luton's straw-hat industry as the town's major employer.

The River Lea enters a culvert under Luton.

The Icknield Way, an ancient trackway crossing Bedfordshire.

Leagrave Marsh, North of Luton, the source of the River Lea.

Onward the River Lea continues until it reaches the grounds of Luton Hoo where it becomes a lake as part of the estate's landscaped gardens which surround the house. The parkland and walled garden of Luton Hoo were designed by the famous landscape gardener Lancelot Capability Brown. In 1763, Brown received his commission from the then resident, John Stewart, 3rd Earl of Bute who became Prime Minister for a short time (1762–3) during the reign of George III. Bute holds the distinction of being the first PM to have come from Scotland. During the Second World War, trials of the now famous Churchill tank, designed and manufactured by Vauxhall Motors in Luton, were carried out within the grounds of Luton Hoo. Perhaps not what Brown had planned for his iconic landscape!

Leaving Luton Hoo behind, the river passes the sites of ancient mills. Then, I imagine the Lea holding its breath as it quickly slips past a sewage works, before crossing the county boundary from Bedfordshire into Hertfordshire at East Hyde. Onward the river flows through the grounds of Hyde Mill Farm, where the present owners have restored the nineteenth-century mill and sensitively converted the farmhouse and outbuildings into bed and breakfast accommodation. The river carries on through the rather large village of

Churchill tank, designed and built at the Vauxhall factory in the Second World War.

Harpenden as it begins its turn eastwards, skirting the edge of Weathampstead, a much smaller village with an interesting early history. Devil's Dyke, a massive earthwork thought to have been constructed in the latter part of the Iron Age (50–100BC), was investigated by the famous archaeologist Sir Mortimer Wheeler in the years 1932-3. Another earthwork feature known as 'The Moat' or 'The Slad' runs parallel some five hundred metres to the east of the site; this, Wheeler suggested, formed the principal defences of an 'oppidum' or 'tribal centre'. It is thought that this centre may have been the headquarters of the British King, or Tribal Chieftain, Cassivellaunus, and could have been the place of his defeat in 54 BC by Julius Caesar.

On leaving the village, the river turns southward and passes through the grounds of Brocket Hall, now a golf and country club licensed for civil weddings and once the home of William Lamb, the second Viscount Melbourne, Prime Minister (1835–41). When Victoria came to the throne in 1837, Melbourne became the Queen's first Prime Minister and confidant. In 1981, the third Lord Brocket turned the house into a conference centre which hosted several government summits. These were attended by Presidents Gorbachev, Regan and George Bush (senior).

As the river leaves the grounds of Brocket Hall, part of the water forms the race for Lemsford Mill, built in 1863. The mill is now the headquarters of a holiday company, Ramblers Holidays. In 2005, the company commissioned Hydrowatt, a German engineering company, to provide an eco-friendly energy source to power the mill building which had been converted into office space. A special Brest-shot wheel was designed to replace the original wooden waterwheel. A drive shaft couples the new wheel to the necessary set of gearing to drive an electric generator. This River Lea powered electricity generating system was the first of its kind to be installed in Britain and can provide sufficient energy for the whole building. Surplus energy is sold to electricity providers.

For many years, historians and others have suggested that Lemsford Mill was the inspiration for the old music-hall song, 'There's an Old Mill by the Stream – Nellie Dean', written by J.P.S. Skelly when he allegedly stayed at Brocket

Lemsford Mill, once thought to be the mill in the old music-hall song, 'Nellie Dean'.

Hall. However, subsequent research suggests that the song was the work of the American songwriter, Harry Armstrong, in 1905. The song became a favourite with theatre goers in the early twentieth century and was made popular by the British music-hall artist Gertie Gitana, whose name is immortalised in Cockney rhyming slang for banana!

After receiving a little more water as it passes Lemsford springs, the river skirts the south-western perimeter of Welwyn Garden City and fills the two man-made lakes in Stanborough Park, opened in 1970 to provide recreational facilities for the nearby town. Welwyn Garden City was fashioned by the visionary creator of the garden city movement, Sir Ebenezer Howard OBE. This was the second of Howard's developments, the first being Letchworth Garden City. Howard held the utopian view that people should have the opportunity to live harmoniously with nature; something that we are just beginning to understand today. For a man born in January 1850, this was really futuristic thinking.

From Stanborough Lakes, the river makes its way to Mill Green Mill, Hatfield, a fully restored eighteenth-century working watermill that is managed by Welwyn and Hatfield Council. The mill grinds organic flour which the resident miller supplies to local bakeries and visitors to the mill can also buy small bags of the product. A museum of local history and art is located in the miller's house.

Mill Green Mill, Hatfield.

On leaving the mill, the tail stream joins the River Lea as it passes under the Hertford Road and enters the wooded grounds of Hatfield House. Here the river enters a linear lake called *The Broadwater* that takes the water in an easterly direction past the now derelict Cecil Saw Mill before exiting the estate. Hatfield House was built in 1611 for Robert Cecil, the first Earl of Salisbury, a trusted Chief Minister of King James I. Within the grounds of Hatfield House stands part of the old wing of the royal palace of Hatfield built in 1497. In the reign of King Henry VIII the palace was claimed by the monarch and this is where his children, King Edward VI and Queen Elizabeth I, spent their formative years.

In 1916, during the Great War, the fourth Marquess of Salisbury, James Herbert Gascoyne Cecil, gave over the grounds of Hatfield House to the military to

De Havilland Comet in BOAC colours built at Hatfield, Hertfordshire.

Hatfield House and grounds where First World War tank trials took place.

Hatfield House and grounds.

allow trials of the world's first fighting tank to take place. Before the vehicles arrived, the estate was turned into a miniature Western Front, with craters, barbed wire and trenches to make the trials authentic. Of course, the town of Hatfield has other historic connections. In the early 1930s, Captain Sir Geoffrey de Havilland built a new aircraft factory in the town where many famous aircraft were made. These included the Second World War Mosquito, known as the 'wooden wonder', and also the post-war Comet, the world's first passenger jet airliner that flew in British Overseas Airways (BOAC) and other airline colours. The site where the factory once stood is now the de Havilland Campus of the University of Hertfordshire.

Stone in the grounds of Hertford Castle commemorating the national church synod of 673 AD.

When leaving the grounds of Hatfield House, the river continues on an easterly cross-country route until it reaches the county town of Hertford. Interestingly, the County of Hertfordshire is represented by the heraldic symbol of two rampant harts (male deer) supporting a shield with blue wavy lines on a white background representing water. Therefore the town of Hertford could have derived its name from harts or a hart fording the River Lea. Hertford is steeped in history which cannot be fully explored here in a few short paragraphs, but I shall give a few examples to whet the appetite of the reader.

In 673 AD, the first national church synod gathered in what are now the grounds of Hertford Castle. Theodore of Tarsus, the eighth Archbishop of Canterbury, appointed by the Pope, united the Celtic Christians of the north with the church of the south. A large stone block within the grounds of the castle records the event. It is claimed that Edward the Elder, King Alfred's son, had wooden fortifications built at Hertford after the River Lea had become the boundary between Wessex, the land of the Saxons, and the Danelaw. This division of land had been negotiated earlier by Edward's father. After the conquest of Britain, a substantial castle was built by the Normans with stone walls and a motte (mound). In the mid-1600s, during the plague years, parliament met at Hertford Castle.

Hertford Castle.

Hertford Lock, where the Lee Navigation begins.

The Lee Navigation begins its journey to the River Thames from the south side of Hertford where the first of eighteen pound locks was constructed. John Smeaton and Thomas Yeomans (both civil engineers) had proposed, in a report of 1776, that this particular section of the River Lea would be suitable as a Navigation after having meanders and flash locks removed; the latter being replaced by pound locks. (Pound locks are controlled by gates at either end; flash locks normally have a single gate that is lifted to release the water.) Now the waterway could bring shipments of grain and malt to the capital in greater quantities and at a swifter pace than had been experienced before when the main means of transport was wagon and horses.

New Gauge on the New River, where water is abstracted from the River Lea.

On reaching the northern outskirts of Ware, the River Lea has to contribute some of its water to recharge the New River, first dug between 1609 and 1613, to take clean drinking water from the Chadwell and Amwell springs to supply London. However, these springs had a tendency to dry up in summertime and, by the middle of the nineteenth century, this supply was insufficient to quench the needs of a growing London population; so it became necessary to construct a system that would abstract just enough water for the capital without hindering the traffic on the Lee Navigation or the work of the downstream millers. The New Gauge building was constructed in 1856 to contain what is effectively a very large ball-cock type mechanism to take on this particular task.[15]

It is believed that the town of Ware took its name from the Saxon word for a weir. In Roman times, Ware was an important inland port. In the eighteenth and nineteenth centuries, the town became famous for its many maltings that kept the brewing industry supplied with a most essential product. Now the town's major employer is Glaxo Smith Klein, the internationally famous pharmaceutical company. According to the Anglo-Saxon Chronicle (written between the ninth and tenth centuries) we are told that in 894 AD a large Viking fleet sailed up the Thames and then up the River Lea to a point twenty miles from London where a fortified camp was built with ditches and

15 For more about how this was achieved, see the chapter on the New River.

ramparts. That would have put the invaders fairly close to Ware. At the time of writing, the author has not been able to discover any convincing archaeological or other evidence to support the story; but this must not be taken as a reason not to continue to look!

As the Lee Navigation leaves the town, it passes, on the east side, the last few remaining eighteenth-century gazebos which were the summerhouses of the rich.

Continuing southwards, the Lee Navigation reaches Rye House just before being joined by the River Stort and Navigation that now accompanies the Lea on its journey to the Thames. Apart from being famous for the Rye House Plot of 1683, there is a more recent story to be told of Rye House. In 1936, during the speedway craze, a motorcycle speedway club was established here which currently supports two teams, the Rockets and the Cobras. However, an even more recent story connected with speed is that of Lewis Hamilton. In 1993, after being bought a Go-Kart by his father, the eight-year-old Lewis began his racing career at Rye House Kart Track and has now become one of the world's top Formula One (F1) racing drivers.

After leaving Rye House, the River Lea and the Lee Navigation pass through an area of the Lee Valley Regional Park's wettest landscape. Consulting a map, it will be noticed that the area is populated with ponds and lakes. These are not natural features but are man-made, left by the extraction of gravel. They occur just north of the Essex town of Waltham Abbey, famed for explosive making, propellant research and much more. In 2012, the town welcomed top Olympic athletes who were competing at the nearby Lee Valley White Water Centre.

The Abbey Church of Waltham Holy Cross and St Lawrence, Waltham Abbey.

White-water rafting near Waltham Abbey.

Royal Gunpowder Mills, Waltham Abbey showing part of the on-site canal system.

As it journeys further downstream from Waltham Abbey, the river leaves behind the county of Hertfordshire and now defines the old boundaries of Essex to the east and Middlesex to the west as it begins to enter what was once the industrial power house of Greater London. Reaching Enfield Lock, the Lee Navigation passes the cottages in Government Row which provided accommodation for the workmen and their families at the Royal Small Arms

Factory (RSAF), the former home of the Lee-Enfield Rifle. After the factory closed in 1987, the site eventually became Enfield Island Village which now, with a commercial hub, supports a vibrant local community. A few hundred metres south of the former RSAF site, the River Lea, via a small manmade channel, supplies water to the most northerly of the Lea Valley reservoir chain, the King George V, opened in March 1913 by His Majesty. The adjacent brick-built pumping station was initially fitted with a set of unique gas-operated pumps that have no piston, mechanical connection devices or flywheel. Herbert Alfred Humphrey, a gifted engineer, designed the pump and patented the system in 1906. Electric pumps are now used to transfer water from the River Lea to the reservoir.

Government Row, former Royal Small Arms Factory workers' cottages, Enfield Lock.

The River Lea continues its journey southward along the east side of the King George V reservoir while the Lee Navigation continues in the same direction on the west side. At Ponders End, a head-stream for Wright's Mill is taken from the Lee Navigation and passes through a small industrial estate beside Duck Lees Lane, a place that I have often claimed as the centre of the universe. In 1904, at the former Ediswan factory, Professor Ambrose Fleming, while investigating a blackening effect within Joseph Wilson Swan's early light bulbs, by accident, invented the diode valve, the world's first thermionic device. This is the first time that engineers and scientists had the ability to control a stream of electrons by electronic means and this allows me to claim that the post-industrial revolution, the technological revolution, had its birthplace at Duck Lees Lane, Ponders End, Enfield. All the electronic equipment that we are familiar with and enjoy today, the radio, the television and the Internet, can be traced back to Fleming's patent of 1904 and thus to a place that is the 'centre of the universe'. Following the redundant head-stream a few hundred metres south, it enters the site of Wright's Mill; its modern equipment is now powered by electricity rather than water. Since at least the Domesday survey of 1086, there has been a mill at Ponders End. The current mill is run by the fifth and sixth generation of the Wright family and is the only commercial working flour mill in Greater London, producing and distributing a staggering range of products to more than satisfy the current popular craze for home baking.

Miller's House, Wright's Mill, Ponders End, home of the Wright family.

THE RIVER LEA – A WATERWAY THAT LINKS US TO THE PAST

On leaving Ponders End, the Lee Navigation continues southward passing the London Waste EcoPark on its west bank before flowing under the North Circular Road (A406) and then through an industrial estate that was once the home of famous furniture manufacturers and their wood yards. Onward the Navigation flows through the Tottenham Marshes towards Tottenham Locks at Ferry Lane, Tottenham Hale. Here, on the south side of Ferry Lane, the Navigation has been widened into a large turning area that was once used to accommodate the barges and lighters that brought wood from the Baltic countries to the Port of London to feed Lebus, the largest furniture factory in the world, and also the other furniture factories that were a feature of this region of the Lea Valley. To people of a certain age, Tottenham Hale is now totally unrecognisable from what they remember from their childhood. The old one-way road system has been replaced with a new road layout, the overground and underground railway stations are now part of a new transport hub that is joined with bus station and taxi rank. A £400m eco-friendly development with rainwater harvesting and communal rooftop gardens, by Lee Valley Estates, includes retail and office space, mixed residential accommodation, a health centre and early learning centre. Running through the centre of the development is a linear park that compliments the biodiversity of the nearby Lee Valley Regional Park.

Barges above Tottenham Lock, Tottenham Hale.

Markfield beam engine, Markfield Park, Tottenham.

As the Lee Navigation leaves Tottenham Hale, it flows past Markfield Park to the west, the present site of a visitor attraction: the Markfield beam-engine museum. This facility is housed in a Grade II listed sewage pumping station with a popular café adjacent. On leaving Markfield Park, the Navigation is still separated from its sister waterway, the River Lea, although it will soon join with its estranged relation once more.

It will be remembered that the River Lea had its original course diverted after leaving Enfield Lock when it was given a subsidiary channel to feed the King George V reservoir. This diversion caused the River Lea to be guided around the Lea Valley's northerly series of reservoirs eventually finding its way to the nearby Walthamstow Wetlands site to the east of the Lee Navigation as it passed under Ferry Lane adjacent to the Ferry Boat Inn. On leaving the wetlands site, the River Lea, now joined by the Coppermill Stream, is reunited with the Navigation in the area of Springfield Marina, a short distance south of Markfield Park.

The combined waterway now continues on its way through the Walthamstow Marshes, designated as a Site of Special Scientific Interest (SSSI), which support over four-hundred different species of insects, plants and animals. The marsh also forms part of the former ancient Lammas Lands that were the meadows on which local parishioners had the common right to graze their animals, from Lammas Day (1st August, the Celtic Midsummer Day) to the following hay harvest on Lady Day (old New Year's Day, 25th March). These rights date back before the Norman Conquest of 1066 and possibly pre-date the Roman occupation. At the southern end of the marsh, a railway viaduct crosses the river carrying commuter trains between Liverpool Street Station and North Chingford (the railway was once owned by the Great Eastern Railway Company (GER) when steam powered the engines). Here, under the brick arches of the viaduct, Alliott Verdon Roe assembled the triplane he had designed, just before he became the first British pilot in a British-built aircraft with a British engine to fly. The engine was designed and manufactured across the River Lea in Tottenham by J.A. Prestwich (JAP). Roe's first official flight was achieved on 13th July 1909 after many failed attempts to get airborne from the marsh. A plaque on the railway viaduct commemorates this historic occasion.

On leaving the marshes, the river flows under Lea Bridge Road where it once provided water for the East London Waterworks Company that had filter beds and a pumping station on the east side. (The story of the East London Waterworks Company and its connections to the Walthamstow Wetlands project will be told later in this book.) Now the site has been transformed into the Middlesex Filter Beds Nature Reserve and the Waterworks Golf Course and Nature Reserve. At this point, the river and the Lee Navigation part company again as the River Lea takes a southward route around the east side of Hackney Marsh while the Lee Navigation continues to the west of the marsh and, for a short period, becomes the Hackney Cut. Prior to the 2012 London Olympics, the River Lea joined with the Navigation again just north of Old Ford locks, but if recent maps are consulted it will be noted that this link no longer exists as the waterway was diverted to accommodate the site for the games.

Walthamstow Marsh, Lammas Lands.

After passing under the East Cross Route (A12), both the River Lea and the Lee Navigation enter the Queen Elizabeth Olympic Park, the river to the east and the Navigation to the west. At Hackney Wick, the Navigation, while still flowing south, is joined with the Hertford Union Canal to the west. This canal was dug in the 1820s to provide a short cut for barge traffic allowing the boats to avoid east London and lighten congestion in the dock area. The Hertford Union Canal now joins with the Regent's Canal, which in turn is linked to the Grand Union Canal, a connection that allows the Hertford Union Canal access to the goods-carrying canal network that links London with the industries of the Midlands and beyond.

Before the River Lea and the Lee Navigation pass their separated ways under the road complex at Stratford, known as the Bow Interchange, the river joins a waterway system, called, in recent years, the Bow Back Rivers. It is believed that the first of these channels was dug in the twelfth century to help drain the area of land known as the Stratford Marsh. As commerce and industry grew, other channels were added and given appropriate names such as: Pudding Mill River, City Mill River, Channelsea River, Three Mills River, Three Mills Wall River and Waterworks River. While the southward-flowing River Lea negotiates the Bow Back River system, the Navigation slips past Old Ford locks to the west. The name 'Old Ford' gives the clue that this part of the waterway was once a crossing point. When the Romans occupied Britain, they

Bazalgette's Abbey Mills Pumping Station, Stratford.

built a road that aligned with the ford then continued onward from Londinium (London) through Essex to their provincial capital, Camulodunum (Colchester).

The Lee Navigation combines, via a short channel, behind the sluice gates of the eighteenth-century House Mill at Bromley-by-Bow, with the River Lea (now called the Three Mills Walled River). Over the years the various rivers that made up the Bow Back River system were given different names to associate them with specific mills or other facilities that they were feeding. When the River Lea entered the Bow Back Rivers to the north of the Three Mills site it effectively became all the other rivers. Hopefully this explanation will allay any confusion in the mind of the reader who has troubled to consult a map of the Lower Lea Valley.

Today only two mills survive, but at the time of the Domesday Survey it is recorded that in the manors of East and West Ham there were eight mills, formerly nine. The House Mill and its partner, the nineteenth-century Clock Mill, are both tide mills that once were worked by the power of the incoming tides that were driven up the River Thames by the North Sea.

Three Mills, Bromley by Bow.

At the time of the First World War, the Clock Mill was being used by Nicholson's as a gin distillery. This is where Chaim Weizmann, who became the first President of the State of Israel in 1948, worked to perfect his process to distil acetone from grain. This ingredient was desperately needed in the manufacture of explosives in support of the war effort. Weizmann asked no reward from the British government for his process but the Minister of Munitions, Lloyd George, wished to grant him an honour. Lloyd George in his memoirs refers to his conversation with Weizmann thus:

I said to him; "You have rendered great service to the State and I should like to ask the Prime Minister to recommend you to His Majesty for some honour". He said; "there is nothing that I want for myself". "But is there nothing we can do as recognition of your valuable assistance to the country"? He replied; "Yes I would like to do something for my people". He then explained his aspirations as to the repatriation of the Jews to the sacred land they had made famous. That was the font and origin of the famous declaration about the National Home for the Jews in Palestine.

Adjacent to the Three Mills site is the Prescott Channel, constructed in the 1930s to take flood water away from the Three Mills. In preparation for the construction of the 2012 Olympic site, a new lock and sluices were built on the Prescott Channel to allow heavy barges to move material to, and spoil from, the site. However, the £23m facility appears to have been scarcely used.

Three Mills Lock on the Prescott Channel, built to service the 2012 Olympics.

The Lee Navigation passes the House Mill to the west and continues southward to Bow Locks where it combines with the River Lea which has now become Bow Creek. Just below the locks, a channel known as the Limehouse Cut leaves Bow Creek to the south west while the main waterway flows a short distance to complete its journey when it joins the River Thames at Trinity Buoy Wharf. This is the site on the west bank of the River Lea where the famous scientist Michael Faraday, in the mid-nineteenth century, oversaw the first experiments in the world to illuminate a lighthouse lantern by electricity. On the opposite bank once stood the Thames Ironworks which built HMS Warrior that, in 1860, was the largest sea-going iron-clad warship in the world. The yard became famous for building many more ships and supplying ironwork for several bridges, tunnels and buildings. The crossed hammers emblem of West Ham United Football Club is a graphic reminder of the club's origins.

This now completes our voyage along the River Lea where I have tried to give glimpses of the many interesting sites that the river passes on its fifty-eight-mile journey. These are some of the footprints that history has left on the Lea Valley's landscape that was shaped by ice sheets receding and melt-waters scouring over ten-thousand years ago. Of course, there are many more places of interest that I could have mentioned but it was considered that by including them it would have been too much of a distraction from the story of the waterway. Should the reader require further information regarding other Lea Valley stories, there are many publications available.

Trinity Buoy Wharf where Michael Faraday carried out experiments to illuminate a lighthouse lantern by electricity.

REFERENCES

Hatts, Leigh, *The Lea Valley Walk*, Cicerone Press, Cumbria, 2001

Lewis, Jim, *London's Lea Valley: Britain's Best Kept Secret*, Phillimore & Co. Ltd, Chichester, 1999

Lewis, Jim, *London's Lea Valley: More Secrets Revealed*, Phillimore & Co. Ltd, Chichester, 2001

Lewis, Jim, Research carried out for a radio series with BBC Three Counties Radio, 2009

HISTORY ON YOUR DOORSTEP – IN AND AROUND THE LEE VALLEY REGIONAL PARK

By now it has probably occurred to the reader that many exciting discoveries have been made and a number of extraordinary events have happened in the Lea Valley. Over the years, historians have generally not paid too much attention to the region so there remains a considerable amount of history to be uncovered and, of course, written up. In the first chapter of this book, we travelled from the valley's early beginnings up to the Norman period and in making this journey it is probable that further questions have been raised in the mind of the reader. Should this be the case, and should the reader be sufficiently stimulated to continue the journey, might I suggest a walk or a cycle ride through the Lea Valley and its Regional Park. While carrying out this enjoyable exercise, it is possible to view some of the places and things of interest that are already known and above ground. Later on, when more familiar with the area, one might like to think about making more interesting discoveries that will add to our store of local knowledge. This could be a very rewarding project, but I must warn you, it can become addictive and time consuming.

While much of my work has tended to concentrate on technological achievements within the region (a subject that had been overlooked by many historians), the Lea Valley has a rich historical heritage to suit all tastes whether those of an experienced researcher or just someone with a casual Sunday afternoon interest. Having such a rich historical and environmental resource on our doorstep also provides some wonderful opportunities for school projects and I would therefore invite teachers to contact the Lee Valley Regional Park's excellent educational services that are located at several of the park's visitor attractions. Information can be obtained from Myddelton House or the Lee Valley Regional Authority's website. One recently formed body, the Lea Valley Alliance, is a grouping of volunteers, museums and other organisations that have come together to promote the Lea Valley's heritage and attractions (at the time of writing, a website is being constructed). The Lower Lea Project, located at the Miller's House, Three Mill Lane, Bromley by Bow, is also an excellent resource for young people and schools wishing to investigate their heritage and to learn about the local environment; and of course, the Lea Valley has a number of superb museums and local history units that are just waiting to give up their secrets. Perhaps the best way of whetting the reader's appetite is by highlighting a few interesting events and places to visit which will hopefully inspire a new generation of young Lea Valley detectives.

Bruce Castle Museum, Tottenham.

Vestry House Museum, Walthamstow.

*Lowewood Museum,
Hoddesdon, Hertfordshire.*

*Epping Forest District Museum,
Sun Street, Waltham Abbey.*

The town of Waltham Abbey is steeped in history. In the 1980s, Roman building materials were uncovered below the Market Square and it has been claimed that King Harold, of Battle of Hastings fame, was buried there (Epping Forest District Museum, Sun Street, Waltham Abbey is a good starting point for information). This Essex town is also the site of Britain's premier gunpowder factory, the Royal Gunpowder Mills, acquired by John Walton and improved by the celebrated engineer, John Smeaton, in the early eighteenth century. The mid-twentieth century saw the creation, out of the gunpowder mills, of the Royal Armament Research and Development Establishment (RARDE) where much important scientific work was carried out. This included the development of high-grade explosives and propellants. In a way, the work was a continuation of the nineteenth-century experimental work, with gunpowder rockets, by the second Sir William Congreve who was also responsible for many improvements at the Waltham Abbey Gunpowder Mills. Congreve's rockets were used on a number of occasions during the Napoleonic Wars and are commemorated in the American national anthem by the wording "the rocket's red glare". This refers to the attack, by the British, on Fort McHenry in 1814, during the Anglo-American War of 1812–14. In the early nineteenth century, Congreve had a factory at Bromley by Bow (currently the site of the old Imperial Gasworks) where he manufactured his gunpowder rockets.

*Large Machine Room at the former Royal Small Arms
Factory site, Enfield Lock.*

*William Congreve II, developer of the
gunpowder rocket.*

A fine example of early English brickwork has been bequeathed to us in the form of Rye House, where the builders used over forty different types of moulded bricks in its construction. Built in 1443 for Sir Andrew Ogard, only the gatehouse survives as a monument to the highly skilled bricklayers of the late middle ages. Rye House once formed part of a large manor, complete with stables and malthouse that stood close to the River Lea; its former grounds are now part of the Lee Valley Regional Park. The house is probably better known for the Rye House Plot of 1683, when the occupier, Richard Rumbold, conspired with others to assassinate Charles II and his brother James, Duke of York, on their return from Newmarket. The plot failed. Two of the conspirators, Lord Russell and Algernon Sydney were executed, the Earl of Essex committed suicide in the Tower of London, Lord Grey and the Earl of Monmouth went into hiding and Rumbold fled to safety in Holland.

The site of Three Mills at Bromley by Bow has a history that goes back to at least the eleventh century. If the reader is not familiar with the area, a visit is highly recommended to take in the cobbled courtyard with its two-hundred-year-old granite tramway designed to give grain and flour carts a smooth passage to and from the mills. I would also suggest that the reader might like to take more than a passing interest in the restored Grade I and Grade II listed buildings and to generally soak up the atmosphere of this historic place which is now a conservation area.

Flagstone tramways at Three Mills, where heavy wagons once ran.

An aerial view of Three Mills showing the island and waterways.

Old Bow Bridge close to the Bow Porcelain Factory; in the background are the chimneys of Bow Brewery.

H. Forman & Son, smoked salmon facility, Hackney Wick.

Abbey Gardens, Stratford, a community garden on the site of Stratford Langthorne Abbey.

Less than a mile downstream from Three Mills is the East India Dock. Completed in 1806, the dock was built on the Thames at Blackwall where East Indiamen had previously moored for almost two hundred years. These ships were famed for sailing to the tropics and returning with their precious cargoes of tea, porcelain, spices, silk and saltpetre. The basin now forms part of a Lee Valley Regional Park nature reserve and bird sanctuary.

Myddelton House at Bulls Cross, now the headquarters of the Lee Valley Regional Park Authority, is named after Sir Hugh Myddelton, who was responsible, in 1609, for successfully raising the capital to begin the construction of the New River. The project, completed in 1613, was designed to bring fresh water from the springs at Amwell and Chadwell, Hertfordshire, to Islington in London. Originally, the New River flowed through the grounds of Myddelton House, built in 1818 and occupied by the Bowles family. It was Edward Augustus Bowles (1865–1954), who, in the late nineteenth century, began the design and layout of the now famous Myddelton House gardens. Gussie, as Bowles was affectionately known, had never received formal training in gardening, but his love and enthusiasm for plants led him to a fifty-seven-year membership of the Royal Horticultural Society (RHA), where he chaired several important committees. E.A. Bowles is probably best remembered for his specialisation in the growing of crocuses and the collecting and propagating of new plant species. For this work and his services to the RHA, he received many prestigious awards and honours.

Phoenix Rose garden centre, Crews Hill, Enfield.

Myddelton House and Gardens, HQ of the Lee Valley Regional Park Authority.

Bowles collection at Myddelton house: a fossil tree.

The Greenway that runs along the line of the Northern Outfall sewer.

Water, that sustains life and is one of the most powerful forces on earth, can be found in abundance in the Lea Valley. This particular resource was put to good use in the nineteenth century by providing the citizens of London with clean drinking water. The supply of clean water, along with a major scheme for the disposal and treatment of sewage, designed by the Lea Valley man Sir Joseph Bazalgette, helped to eradicate the typhoid and cholera epidemics that had claimed the lives of thousands of Londoners.

A short distance north of the Middlesex Filter Beds, Lea Bridge Road (now a nature reserve) near where the Liverpool Street to Chingford railway viaduct crosses the River Lea and Walthamstow Marshes, an event took place that was to set the scene for Britain's aviation industry. As mentioned in the last chapter, on the 13th July 1909, Alliott Verdon Roe made his historic flight in a triplane powered by a 9hp J.A. Prestwich (J.A.P.) engine, manufactured just across the River Lea in Tottenham. This was the first all-British flight of a British aircraft with a British engine and flown by a British pilot. A blue plaque, placed on the viaduct wall adjacent to the railway arch where the plane was kept and prepared, marks the location of A.V. Roe's flight.

These are just a few examples of the hundreds of interesting events that have taken place in the Lea Valley region. To learn more, the reader is invited to refer to the list of Lea Valley books on the cover which are to be found at some local authority history units and libraries.

Hopefully, by highlighting these few examples along with the other subjects uncovered in this book, it will have been demonstrated that the Lea Valley has a fascinating past with many more secrets to be uncovered. Who will become the next Lea Valley Sherlock Holmes?

REFERENCES

Ashby, Margaret, *The Book of the River Lea*, Barracuda Books Ltd, 1991

Author unknown, *Rye House Revealed*, Lee Valley Regional Park Authority, 1997

Bascombe, K.N., and Bentley, John, *A Walk Round Waltham Abbey*, Waltham Abbey Historical Society, 1998

Boyes, J., Smith, D., Pearson, A., Whitehead, W., Gray, M., Fairclough, K., and Wurzell, B., *A Guide to the History and Wildlife of the Middlesex Filter Beds Nature Reserve*, Lee Valley Regional Park Authority, 1991

Burnby, J.G.L., and Parker, M., *The Navigation of the River Lee (1190-1790)*, Edmonton Hundred Historical Society, Occasional Paper New Series No.36

Camp, John, and Dean, Dinah, *King Harold's Town*, Waltham Abbey Historical Society, 1988

Field, Tania, and Sainsbury, Frank, *Stratford's Hidden Heritage*, Lee Valley Regional Park Authority, 1995

Fraser, Antonia, *King Charles II*, Weidenfeld and Nicolson Ltd, 1993

Heath, Cyril, *The Book of Amwell*, Barracuda Books Ltd, 1980

Hewitt, Bryan, *The Crocus King - E.A. Bowles of Myddelton House*, The Rockingham Press, 1997

Higgins, Eric, and Bascombe, Kenneth, *Waltham Abbey*, Pitkin Pictorial Ltd, 1995

Lewis, Jim, *London's Lea Valley: Britain's Best Kept Secret*, Phillimore & Co. Ltd, 1999

THE NEW RIVER

An imaginative scheme to bring fresh drinking water to the City of London from the springs situated north of the metropolis at Amwell and Chadwell in Hertfordshire, a distance of approximately twenty miles, began on 21st April 1609. After several acts of parliament concerned with bringing water to London and a rather dilatory approach to such schemes by the Corporation, the high cost seemingly slowing progress, a proposal put forward by Hugh Myddelton (later Sir) to fund and manage the venture was accepted formally (perhaps enthusiastically) on 28th March 1609.

Sir Hugh Myddelton who, with King James I, financed construction of the New River.

Myddelton was born at Galch Hill near Denbigh, in or about the year 1555. He came from a large family of nine brothers and seven sisters. His father, Richard, who had been Governor of Denbigh Castle, died in 1575. His mother Jane had departed this life some ten years earlier in 1565. Hugh started his working career apprenticed to the Goldsmiths' Company and it would seem that part of his employment included banking and money-changing. Between the years 1603 and 1628, he had been returned as Member of Parliament for the Borough of Denbigh on six occasions. It is probably fair to say that, had Hugh lived today, he would be labelled a venture capitalist and risk taker.

The plan for the construction of the New River, as it became known, was to follow closely the hundred-foot contour along the western slopes of the Lea Valley, the destination being a storage pond or reservoir which was to be dug at Islington. By deciding this route for the river, it effectively doubled the distance the water had to travel from twenty to almost forty miles. However, what is so staggering about this project, which must be seen in the context of the day as a considerable feat of engineering, is that the channel dug from Hertfordshire brought water to the city by gravity alone. The average fall of water to Islington was only 5.5 inches (14cm) per mile. In Britain, at that time, the use of pumps for shifting water was relatively rare.

Although construction of the New River was held up for almost two years due to disputes with landowners over the amount of compensation to be paid, the work was completed when the course was extended to Islington in April 1613. The official opening ceremony took place on 29th September that year with much celebration. When considering the problems which had to be overcome, the speed of the waterway's completion was truly remarkable. It is recorded that 157 bridges spanned the length of the river, which generally flowed north to south. However, there were many streams to negotiate which ran west to east across the valley carrying land drainage water. To reduce the risk of possible contamination to the clean water in the New River, these streams were allowed to follow their natural course towards the River Lea by being taken beneath the line of the new channel.

Chadwell springs, north of Ware.

An early drawing of New River Head at Islington showing filter beds and reservoirs where the waterway originally terminated.

When the New River was dug, Islington was effectively a village set in open country and situated approximately one hundred feet above the level of the Thames. The site of the storage pond was chosen quite deliberately to take advantage of the natural fall of the land which gently sloped towards the city. This made water distribution relatively easy as it was possible, under gravity, to pipe water to the height of the second floor of some houses.

By today's standards, water distribution was rather crude. The main conduits were positioned above ground, sometimes on trestles, and were constructed from drilled sections of elm-tree trunks. Each section was joined to the next by creating a friction joint secured with an iron ring. One end of a section was made to fit the next by shaping the mating piece like the sharpened end of a pencil. Individual house supplies were taken from the wooden main via a small-bore lead pipe which was usually terminated with a swan-necked cock for drawing off the water.

While those who benefited from this new method of water distribution were no doubt overjoyed, the system was very inefficient. Early in the nineteenth century, it was reported that there were losses from the supply of twenty-five percent, attributed to leaks from pipes alone. However, it would be unwise to be too critical, considering the primitive nature of this early technology, as it has been reported in recent years that losses of water from leaking pipes in some regions of Britain have been as high as forty percent. As the population of London increased, so did the demand for fresh drinking water. Over the years the New River saw many modifications to increase and quicken the flow into London. Bends were straightened, wells were dug, reservoirs were built and pumps installed. Today, when trying to trace the original route of the New River it will be noticed that much has been filled in. However, due to sustained pressure from environmentally conscious community groups, a considerable amount of this ancient waterway has been saved.

Today, some of the remaining sections of the New River form an integral and important part of a much larger and complex system of reservoirs, treatment works, pumping stations and filter beds stretching along much of the length of the Lea Valley. In recent years a scheme was completed to take water from the New River to recharge the region's depleted aquifer, effectively

Broadmead Pumping Station, built in 1880 at Ware, now uses modern electric pumps to lift water from the aquifer to recharge the much-straightened New River.

Woodbury Wetlands nature reserve on the New River reservoirs, Stoke Newington; the New River now connects to the Coppermill Waterworks which feeds the London Ring Main.

Built in 1856, the New Gauge regulated the flow of water from the River Lea to recharge the New River.

Chadwell Miln's pumping station at Stoke Newington, now the Castle Climbing Centre.

connecting the ancient artefact to Thames Water's state-of-the-art London Ring Main which tunnels through clay forty metres below the metropolis. Incidentally, the depth of the Ring Main was chosen so as not to interfere with the Transport for London (TfL) underground railway system.

The New River, which used to terminate at New River Head, Rosebury Avenue, Islington, now ends effectively at the east reservoir on the site of William Chadwell Miln's pumping station (now the Castle Climbing Centre) that stands above Stoke Newington in Green Lanes. I say "ends effectively" with a certain amount of caution as this old lady has still an important role to play in keeping us healthy. An underground pipe connects the east reservoir to the Coppermill Stream in east London which runs through the

Walthamstow Wetlands site taking New River water to the Thames Water's Coppermill Lane works and filter beds. So it can now be claimed that, apart from recharging the region's underground aquifers, the New River is directly connected to the London Ring Main.

Who could have imagined that over four-hundred years after the completion of the New River, the citizens of London would still be deriving benefit from the water brought to them by the remarkable skill and achievements of those early engineers and surveyors who designed and planned this waterway? Also, we should not forget the contribution made by the labourers, who through hard manual toil shifted hundreds of tons of earth in digging the course of the river, no doubt helped in their endeavours by the odd horse and cart to take away the spoil.

REFERENCES

Author unknown, *London's Water Supply in the 21st Century: A Strategy for Water Treatment and Trunk Distribution*, published by Thames Water, February 1986

Harwood, Elaine, *The New River*, Report by English Heritage, August 1989

Morris, R.E., *History of the New River*, 1934

THE EAST LONDON WATERWORKS COMPANY

The East London Waterworks Company (ELWC), which has historical connections to the Walthamstow Wetlands, was established under an Act of Parliament in 1807 on a thirty-acre site at Old Ford, Bow towards the southern end of the River Lea. The Act defined the area of east London and Essex to be supplied with water and also authorised the Company to abstract water from the River Lea "at or near Old Ford". However, this was only to be done "during such times as the tide water should be flowing up the river, and had flowed to such a height as to stop the working of the mills below on the river". Over the years there have been many disputes recorded between mill owners, barge owners and other river users regarding the use of the River Lea's water. However, it will be noted that the specific wording of the Act seems to be trying to avoid new tensions arising between users. A further Act of 1808 allowed the ELWC to take over the former Shadwell and West Ham Waterworks and this no doubt considerably increased the dominance of the Company in the region.

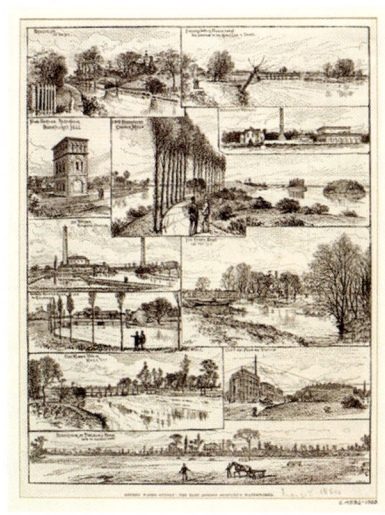

A collection of drawings showing the development of the ELWC site at Lea Bridge.

It has been reported that the opening ceremony of the ELWC on 23rd October 1809 was a rather grand affair and the occasion was probably used to advertise, to the surrounding neighbourhood, the establishment of this new water supplier. We are told that the court of directors first assembled at company headquarters at Bishopsgate and from there they travelled in procession by carriage, led by the chairman Sir Daniel Williams and his deputy John Ord, to the parish of St Mary Stratford, Bow where a special service was held. Afterwards, the directors and their entourage proceeded to the works at Old Ford where a covered platform with seats had been built (apparently the cover had been erected specifically "for the accommodation of the ladies"). Colonel Beaufoy of the 1st Royal Tower Hamlets Militia with his regiment and military band were in attendance. When the director's party was comfortably seated, a bugle call sounded to announce the arrival of the Lord Mayor. Numerous speeches were then made, including one by the Company's engineer, Ralph Walker.

River-god sculpture, originally installed at the ELWC, Old Ford in 1809, then moved to Lea Bridge Waterworks and now at Coppermill Waterworks.

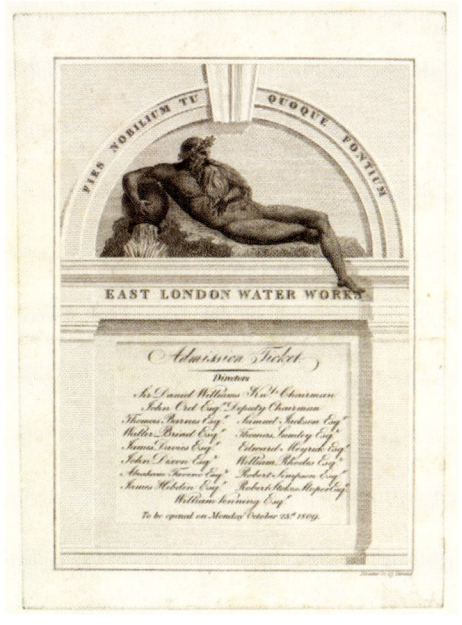

Invitation to the opening of the ELWC at Old Ford in 1809.

After the formalities the Lord Mayor, followed by the gathering, made his way to the site's first reservoir where he ceremonially opened the sluice allowing water from the River Lea to flow into the system. Three hearty cheers went up and the Lord Mayor then proceeded to open the second sluice to the second reservoir. A party of artillery men fired a twenty-one-gun salute and the band played 'God Save the King' followed by 'Rule Britannia'. It would seem that an enjoyable day was had by all and it is my guess that, as on so many of these occasions, the great and the good were whisked away to indulge in a sumptuous lunch!

The directors of the ELWC must have expended considerable sums of money on the planning and delivery of such a grand opening ceremony with apparently little thought for their not-so-fortunate customers, who were about to be supplied with the Company's not-so-pure water. The water was taken from the tidal section of the River Lea (as the 1807 Act had specifically stipulated) and thus came mostly from the polluted River Thames. It was then pumped by steam engines, untreated, from the two settlement reservoirs to the surrounding population through the Company's mains (see the later chapter regarding the negligence of the ELWC). The early legislation had laid down that the various London water companies should compete for their customers but in 1815 the ELWC drew up a legal agreement with the New River Company which would identify their respective areas of supply. Under the circumstances, this would seem like a sensible arrangement and it later became an advantage for the New River Company when an investigation took place into deaths in the area (see later chapter).

Due to increasing industrialisation, the population of London had expanded rapidly as people were drawn to the capital for work. Without strict laws to control the dumping of industrial and other waste, the poor levels of sanitation that already existed in the capital increased. Run-off of effluent into water courses, leaching of cesspits into aquifers, coupled with increasing levels of air pollution all directly affected the cleanliness of drinking water.

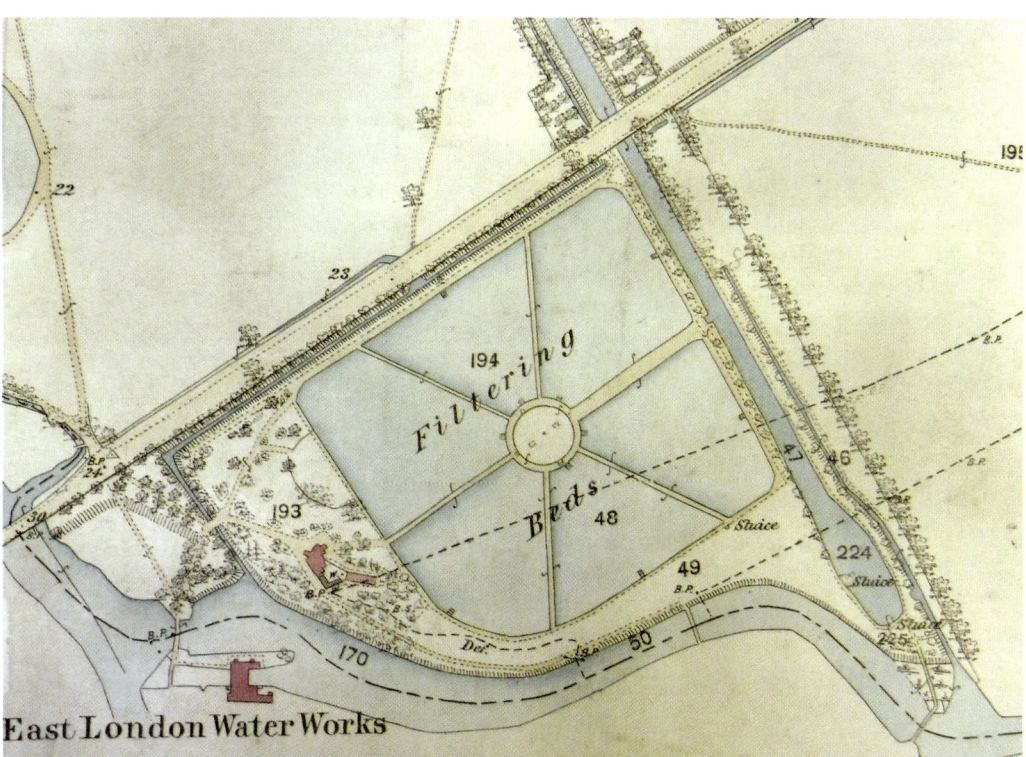

Lea Bridge filter beds, ELWC.

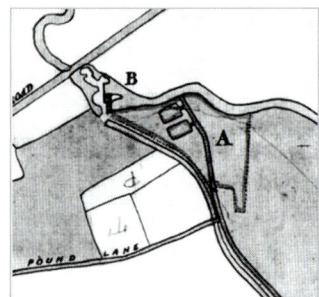

Plan of Lea Bridge c.1828 showing reservoirs with channel to feed waterworks at Old Ford.

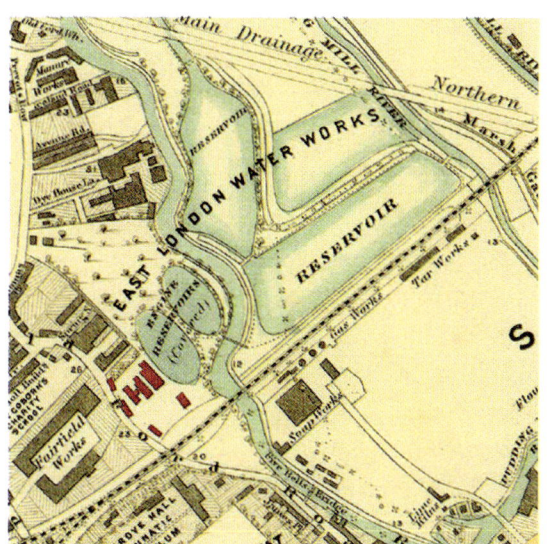

Map of ELWC, 1862, showing first reservoirs west of River Lea and later reservoir complex to the east.

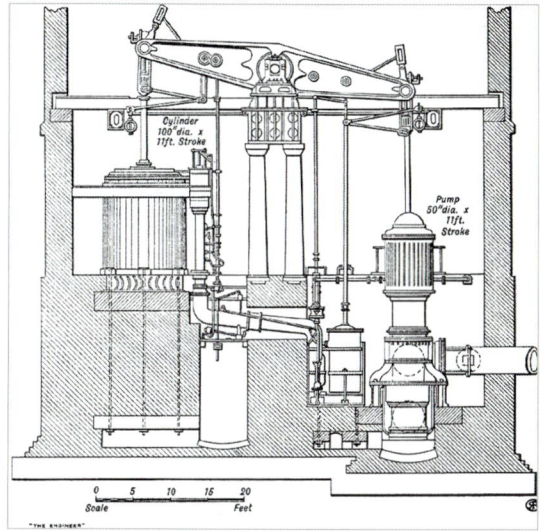

Line drawing of beam engine installed at ELWC Lea Bridge pumping station.

Middlesex Filter Beds and pumping station, 1936.

Not paying attention to these problems would result in a major health epidemic in the capital, the results of which would have serious implications for the ELWC.

By 1829, the ELWC had moved its point of abstraction further north up the River Lea to the Lea Bridge area on the edge of Walthamstow Marsh. In the same year, William Hoof, a specialist contractor in water delivery projects, was commissioned to plan and build a channel adjacent to the Hackney Cut (part of the River Lee Navigation) that would take the abstracted water to the Company's reservoirs at Old Ford. It can be seen from a map of 1843 that two new reservoirs have been dug at Old Ford and a later map of 1862 shows a third. These extra reservoirs would have been necessary to hold water taken from the Lea Bridge channel and their appearance on the maps would suggest that the population of London was growing and demanding more water. As it will be appreciated, taking water from higher up the River Lea overcame the problems of taking the polluted incoming Thames water from the river's tidal reach.

When the Metropolitan Water Act of 1852 came into force it granted the ELWC permission to construct new reservoirs at Lea Bridge and also further north at Walthamstow. It is interesting to note that the Lea Bridge site had seen much earlier industrial usage. In the early eighteenth century a waterwheel was used for pumping water to a reservoir at Clapton where distribution to local houses took place. In 1760, the Hackney Water Works Company was set up and took out a lease on the site. After improvements to the facilities, water was pumped from the River Lea to the surrounding locality. Over the years the Lea Bridge Mills were used to provide power to grind corn, pump water and drive machinery to bore tree trunks for making wooden water pipes. In 1796, a serious fire devastated the Mills, destroying the waterworks and a factory that manufactured pins and needles.

By 1895 the Lea Valley reservoir complex had grown from two to twelve and had a surface area of 479 acres. In 1902 the amalgamation of all the private water companies in London saw the creation of the Metropolitan Water Board (MWB), a public-sector body. Over the following years, investment and modernisation in the region's water complex by the MWB, which later became Thames Water, took place. In 1969, due to a modernisation programme, the Lea Bridge works, known as the Middlesex Filter Beds, were closed. Responsibility for water treatment was transferred to a new purpose-built plant at Coppermill Lane, Walthamstow. Here, water from the Lea Valley supplies around thirty percent of London's needs.

In 1988, after many years of vandalism and neglect, the Lee Valley Regional Park Authority acquired the Lea Bridge site. Local people were consulted and the outcome was the creation of an imaginative scheme to establish a nature reserve out of what remained of the derelict industrial complex. Instead of having the filter beds removed, they became sensitively integrated into the overall nature reserve. Now, after careful husbandry and having

Remaining Lea Bridge Waterworks buildings.

taken only a few years to become established, the reserve can boast more than two-hundred plant species; and over sixty different bird varieties have been recorded there. The reserve is also a haven and a breeding ground for a variety of amphibians, dragonflies, damselflies, butterflies and a host of other insects. In fact the Middlesex Filter Beds Nature Rserve is recognised as one of London's most valuable wildlife resources.

This is a good example of a public body engaging with the local community to create a wildlife sanctuary in an area that, over the years, had seen nature destroyed by industrial development. Now young people and their families have a tranquil place to visit where they can learn how to conserve nature and protect our wildlife which will allow them to pass their newly acquired knowledge to the next generation. By continuing the appreciation and preservation of nature, and understanding how inextricably it is linked to the well-being of our planet, maybe it will be possible to halt the damage we are all doing to the environment by our avaricious quest for more and more goods and services, the production and development of which adds to our carbon footprint.

REFERENCES

Powell, Andrew B., *Fields and Factories: The Making of the Lower Lea Valley*, Oxbow Books, 2012

Powell, W.R., (ed.), *A History of the County of Essex: Volume 6: West Ham: Rivers, bridges, wharves and docks*, 1973

Lewis, Jim, *Water and Waste: Four Hundred Years of Health Improvements in the Lea Valley*, Middlesex University Press, 2009

TOTTENHAM'S CONTRIBUTION TO PUBLIC HEALTH

It is relatively easy to recognise and appreciate those structures in our community that create the goods and the jobs that sustain our quality of life; but it is also a simple matter to forget those crucial services that were bequeathed to us by our Victorian ancestors that have contributed, and still contribute, to the health and well-being of so many of us today. It is also possible to forget that these structures, goods and services, if not properly designed and developed, can have a negative effect upon our environment. Our planners must therefore reach a compromise.

Some of the services referred to, occasionally considered too delicate to talk about in polite circles, are the plants and buildings responsible for the treatment and disposal of the sewage that we create on a daily basis. Perhaps, when we dutifully pull the chain, turn the handle or push the button to flush the toilet, it is a question of out of sight, out of mind.

It is probably hard to imagine that in the early part of the nineteenth century, around 180 years ago, at the time when Queen Victoria came to the throne, infant mortality in British towns was bordering on fifty percent. Children were dying of conditions that are easily preventable today such as typhoid, diarrhoea, dysentery and cholera. London, in the nineteenth century, was struck by epidemics of both cholera and typhoid that claimed the lives of thousands. The mechanisms which accounted for the spread of these horrific diseases were not fully understood at the time.

While we are aware today of the importance of personal hygiene and the need to have clean and fresh drinking water, Victorian medical science had not generally progressed to this state of understanding. In 1844, a leading chemist, Professor Booth, had considered that it was the "free currents of air" that were circulating from the River Thames that were the cause of London's ailing health. The eminent scientist, Michael Faraday, who recognised the poor state of the Thames, which he described in a letter to the *Times* of July 1855 as a "fermenting sewer", seems more concerned about the pungent smell rather than the real cause of the disease which could be the capital's consumption of untreated river water. The concept of diseases being spread by air pollution, the 'miasmic' theory, appears to have been a fairly common notion for many Victorians. It is claimed that even the famous lady with the lamp, the nurse of the Crimea, Florence Nightingale, went to her grave in 1910 still believing in the theory.

Further evidence of the general acceptance of the miasmic theory amongst some of the most prominent men in the land can be gleaned from the reports of the 'Great Stink' during the very hot summer of 1858. This was the year when the incoming tidal flow of the Thames brought the filthy effluent, which people and local authorities had discharged through the various water channels, and which had also leached from overflowing cesspits into the river, to a place opposite the Houses of Parliament. In an effort to eliminate the dreadful smell that occurred in the House, curtains soaked in chloride of lime were hung over windows and doors, presumably in the hope of protecting the members from the miasmic effect. It has also been reported that the then Chancellor of the Exchequer, Benjamin Disraeli, was seen leaving the House in a hurry with a handkerchief clasped firmly to his nose.

It was probably the Great Stink, more than anything else, that helped to cut through all the bureaucratic layers, delays and financial hold-ups that Joseph Bazalgette, the Chief Engineer to the Metropolitan Board of Works, had endured. In the same year, he was finally allowed to begin the task of building the London sewage system. For its day the project was a truly massive civil engineering undertaking. Not only were 1,300 miles of brick-lined sewers to be completed together with the main northern and southern outfall pumping stations, but also roads had to be redesigned and aligned, bridges improved and the mammoth task of constructing the embankments on either side of the River Thames had to be dealt with. The system had to be designed not just to dispose of London's sewage safely but also to cope with the drainage problem, particularly in times of high rainfall (for more on Joseph Bazalgette, see the relevant later chapter).

Tottenham is a borough that now borders the Walthamstow Wetlands but, by the middle of the nineteenth century, it was suffering the effects of an expanding population. As in other areas of London, this was placing considerable pressure on the antiquated methods of sewage disposal. By the late 1840s it is estimated that around 800 dwellings were discharging their effluent waste into the Moselle Brook, which connected to the River Lea, a main artery that flowed south to the Thames. It would appear that the authorities had identified the problems, as Tottenham became one the first boroughs in Middlesex to take powers under the 1848 Public Health Act to establish, in 1850, a permanent Local Board of Health. The Board consisted of nine members who had financial powers and this allowed them to put in place plans for a public water supply and a system for the disposal of sewage.

A site, adjacent to the River Lea, at the end of Markfield Road, was chosen as the preferred spot for building a sewage works. To deal with the effluent waste, a forty-five-horsepower horizontal steam engine, that drove a single double-acting pump, was installed. Incoming sewage to the works was pumped to a pair of settlement and holding tanks where the liquid effluent was passed through a series of sand filters and eventually discharged into the River Lea. Solids were removed from the site by a local manure contractor who had the responsibility for treatment before selling it on to neighbouring farmers. Two workmen lived on site in a pair of small purpose-built cottages.

A 45hp condensing engine was the first to be installed at Markfield to pump sewage sludge.

Two early engine houses with chimneys at the Markfield sewage treatment works.

Early map showing ownership of land near Tottenham Hale; Marke Fields can be seen centre top.

Their job was to provide twenty-four-hour security, maintain the steam engine and stoke the boiler. By mid-nineteenth-century standards, it could be argued that Tottenham now had a relatively efficient sewage treatment and disposal system.

By 1853, the Local Board of Health claimed to have completed their arrangements to supply clean drinking water to all the built-up areas of Tottenham. However, by 1856, due to increased demand, it was found necessary to extend the waterworks at Tottenham Hale and water was abstracted from the surrounding marshlands. In 1858 a major problem occurred for the Board with the death of the local manure contractor. For some unaccountable reason the sewage solids were not collected and the upshot was that the effluent was allowed to overspill onto the adjacent marshland and some was deliberately discharged into the River Lea. Eventually, the water that was being abstracted by the waterworks became contaminated due to the leaching of sewage through the soil.

In 1866 a typhoid epidemic in east London caused the deaths of almost four thousand people. The East London Waterworks Company (ELWC) had supplied many of these unfortunate residents with water. As we know, the Metropolitan Water Act of 1852 had granted the Company permission to build new reservoirs at Lea Bridge and also further north up the River Lea at Walthamstow. Because of Tottenham's effluent discharges into the river, the ELWC accused the Tottenham authorities of causing the deaths of their customers. While Tottenham's discharges were clearly not in the best interests of public health, the ELWC's accusations could have been a distracting tactic to divert blame away from their own operation.

One of the provisions of the 1852 Metropolitan Water Act was to ensure that reservoirs within a five-mile radius of St Paul's Cathedral should be covered. In September 1866, a group of twenty-nine residents (the Act provided for a minimum of twenty to complain) supplied with water by the ELWC brought a complaint to the Board of Trade, claiming that the water supplied by the Company was in fact contaminated water from the River Lea. A Captain Tyler was appointed to investigate the truth of the matter and found that, on at least three occasions during 1866, the Company was clearly in breach of the 1852 Act as water had been taken from an old uncovered reservoir, that was subject to contamination, and fed to one of their closed reservoirs, used for supplying drinking water. Tyler also concluded that between July and August 1866 an estimated 4,363 deaths had occurred, 3,797 of these being in areas supplied solely by the ELWC. There was a further 264 deaths in an area shared with the New River Company, making the ELWC wholly or partially responsible for over ninety percent of the fatalities. For these reasons, in an earlier chapter, I suggested that the ELWC was negligent in maintaining an adequate safe standard of distribution for its customers. Furthermore, it is recorded that dead and rotting eels had been found in the Company's water pipes, suggesting that even at this late stage water was still being abstracted from the River Lea and passed to customers unprocessed. The agreement reached in 1815 between the ELWC and the New River Company, effectively dividing their respective east-end water customers into two distinct groups, had probably helped Captain Tyler identify the real polluting culprit much more easily.

As the nineteenth century progressed, many east and central London industries relocated to the relatively inexpensive land that was available in Tottenham and other upper Lea Valley areas. This produced an increasing pressure to boost levels of sewage treatment and disposal, as well as the supply of clean drinking water. The expansion of the Great Eastern Railway network into the region with the offer of the cheap workman's ticket and the later introduction of the electric tram made the area popular with commuters who were able to follow the jobs. It was not long before builders, with an eye for business, realised there was a growing market for affordable houses and this in turn applied further pressure on the Local Board of Health.

In 1885, things came to a head with the publication of a highly critical River Lea Conservancy report which stated, among other things, that "the real source of the condition of the Lea, other causes apart, was the neglect of the Tottenham Board... not providing sufficient tank accommodation for their rapidly expanding district". It would therefore appear that in a relatively short space of time, the once forward-thinking Local Board had allowed the provision of public health to descend into chaos.

To address the problems, the Local Board consulted an experienced surveyor who came up with a seven-point plan to expand the Markfield sewage works considerably. However, only two of the surveyor's recommendations were implemented: the installation of a new beam engine and the lengthening of the settlement and holding tanks. By January 1886, the construction of a new engine house was completed and by November of that year Messrs Wood Brothers of Sowerby Bridge, Yorkshire, had installed a 100 horsepower beam engine to drive two single-acting plunger pumps. The efficiency of the engine, commissioned on the 12th July 1888, was such that over a twenty-four-hour period it could pump four million gallons of sewage sludge while turning at only sixteen revolutions per minute. With the availability of the new engine, the powers granted under the 1886 Lee Purification Act and the construction of a new sewer linking into the Hackney Branch Sewer, the Markfield works was now able to discharge the borough's effluent waste directly into Bazalgette's London sewage system for onward transmission to Beckton, rather than having to discharge into the River Lea as had been done in the past. Not surprisingly, within a few years of these new arrangements, the Lea Conservancy reported that "fish are now numerous in the river below the Tottenham Sewage Works".

The short sightedness of the Tottenham Local Board of Health in not implementing all the recommendations of their surveyor meant that, by the early twentieth century, with a rapidly growing population, the whole issue of sewage disposal had to be addressed yet again. In 1905, a second engine house had been completed by a local builder, Rowley Brothers of South Tottenham, in which three Worthington steam engines were installed along with six double-action plunger pumps and eighteen steam cylinders. The system was powered by any of three Lancashire boilers, two old and one new. The works' total pumping capacity increased by seventy-five percent to twenty million gallons in a twenty-four-hour period, allowing it to handle elevated levels of storm water. Under normal running conditions, the plant would be limited to pumping ten million gallons of sewage per day. Because of the increased pumping capacity of the works, an extra sewer had to be laid in parallel with the original. Remarkably, the total cost of all the building work and the installation of all the equipment only came to £43,800, a veritable snip by today's standards.

One of the three Lancashire boilers that provided steam for the Worthington engines.

In 1905, three sets of triple-expansion horizontal Worthington engines were installed to replace the original 45hp engine.

Once the installation of the new plant was complete, the forty-five-horsepower pumping engine, installed in the early 1850s, was scrapped but the faithful Wood Brothers beam engine was retained, remaining on standby in its original engine house in case of emergencies. It was planned only to bring the engine on line if storm water levels appeared to be rising at a dangerous rate.

The installation of the new facilities at Markfield seems to have been eminently timely. Only the year before, Harris Lebus, the furniture maker (who by the late 1940s was to become the largest furniture manufacturer in the world) opened a brand new factory on a 13.5 acre green-field site at Tottenham Hale. As many of Lebus's employees had to travel to work from London's East End, Harris encouraged builders to create affordable homes for his workers locally and this caused considerable expansion of South Tottenham's housing stock.

By the 1930s, because of the vast increases in the development of industrial, commercial and residential buildings across the region, Middlesex County Council had taken responsibility for the treatment of sewage. To address the problems that these increases in building had brought, plans were drawn up for the construction of a new sewage works to the west of the County at Mogden Lane, Isleworth and another in the east at Deephams, then in the borough of Edmonton (now Enfield). Completion of the Mogden Lane works was carried out ahead of the Second World War but Deephams' construction was postponed and not completed until 1963.

Like other areas of London, Tottenham had suffered the nightly raids of the Luftwaffe during the Second World War. Although the engine house was damaged by incendiary and explosive devices during three raids in the autumn of 1940, which mainly caused damage to the roof, the Markfield engines were able to keep working for the duration of the War. However, during this period, apart from providing the essential sewage-treatment facilities, a pig farm with a house for the manager was built on the site as a way of boosting wartime food production. Local residents were encouraged to deposit waste food in specially provided bins located on street corners. The contents of these were collected and processed by the local authority

Queen Elizabeth, the Queen Mother, visiting the pig farm established at Markfield during the Second World War. Pigs were fed with food waste – recycling is not new!

and turned into a source of animal feed that became affectionately known as 'Tottenham Pudding'. So it can be seen that the idea of recycling household waste is not a particularly new phenomenon and when there is an urgent need to do something it can be achieved.

The completion of the Deephams sewage works in 1963 caused the new London Borough of Haringey, now responsible for Markfield, to consider the site's future and the decision was taken to use the new works at Edmonton, located only a few miles to the north. A trunk sewer was dug from the Markfield works to the new facility and it was so arranged that sewage could flow under gravity. This meant that Markfield's pumping provision was now redundant and so, in February 1964, the works were finally closed and the surrounding site levelled. Over the coming years, as might be expected, the sight of lonely buildings in an isolated location inevitably attracted the attention of vandals and the engine house sustained more damage from them than the Luftwaffe had been able to achieve. However, it was not until 1974 that the windows of the engine house were bricked up to protect the historic beam engine inside, and the site's accumulated rubbish removed.

Luckily, a small group of enthusiastic volunteers recognised that something had to be done to protect this part of Tottenham's industrial heritage. They came together and formed what eventually became the Markfield Beam Engine and Museum Ltd, a company limited by guarantee and a registered charity, with the aim of maintaining and renovating the engine and its building. This action has probably prevented the loss of what is believed to be the last engine manufactured by Wood Brothers. It is certainly the only surviving eight-column beam engine still residing in its original location.

The quality of the material used in the Wood Brothers' engine can be gauged by the following single sentence written by G.R. Stephens, the Borough Engineer and Surveyor, in October 1984: "The Worthingtons finally succumbed to red rust (for the first time in their lives!!) and were scrapped". It should be remembered that these Worthington engines were almost twenty years younger than the Wood Brothers' beam engine.

To help secure the future of the engine house, the Museum applied to English Heritage for listing and was eventually awarded Grade II status. Further

Markfield pumping station and community centre today: a good example of recycling buildings for heritage and community use.

applications were made for funding and finally the volunteers were successful in receiving £3m from the Heritage Lottery Fund and other sources which has helped to fully restore Markfield and build an on-site café. Now the Museum's volunteers have achieved their ultimate goal in making the beam engine a working exhibit, thereby creating an exciting visitor attraction.

On special days, visitors are transported back into the past by the sight of the monster engine belching steam, the great horizontal beam rocking on its pivot point and the flywheel spinning smoothly. The Museum also continues to ensure that the site provides a first-class educational facility for young people, particularly groups from local schools who are encouraged to become involved in simple research and other projects related to the site.

The Markfield site is a short walking distance from the Walthamstow Wetlands, which were created from former reservoirs. This gives visitors the opportunity to appreciate how our Victorian ancestors used massive steam engines to pump not only sewage sludge but also vast quantities of water to and from reservoirs and onward via waterworks and filter beds to our houses.

REFERENCES

Author unknown, *The Opening of Outfall Works Extension*, Tottenham and Wood Green Joint Drainage Committee, July 1905

Author unknown, *Fit to Drink*, Walthamstow Antiquarian Society, 1986

Brereton, Kenneth, *The Site History of the Markfield Beam Engine*, Markfield Beam Engine and Museum Ltd, 2007

Clark, Frederick (Dr), Secretary of the Markfield Beam Engine and Museum Ltd, personal conversation, August 2007

Halliday, Stephen, *The Great Stink of London: Sir Joseph Bazalgette and the Cleansing of the Victorian Metropolis*, Stroud, Gloucestershire: Sutton Publishing Ltd, 2001

Hedgecock, Deborah, Curator of Bruce Castle Museum Tottenham, personal conversation, August 2007

Lewis, Jim, *London's Lea Valley: Britain's Best Kept Secret*, Chichester: Phillimore & Co. Ltd, 1999

Lewis, Jim, *East Ham and West Ham Past*, London: Historical Publications, 2004

Lewis, Jim, *Water and Waste: Four Hundred Years of Health Improvements in the Lea Valley*, Middlesex University Press, 2009

Parliamentary Papers, Vol. 58: *Report of Captain Tyler to the Board of Trade, in Regard to the East London Waterworks Company*, 1867

THE LEGACY OF AN ENFIELD MAN

Joseph William Bazalgette (later Sir Joseph), not to be confused with his father, Joseph William Bazalgette, a Commander in the Royal Navy, was born at Enfield, Middlesex. When communicating with his great-great-grandson, the television executive Sir Peter Bazalgette, I learned that Sir Joseph's biographer was unaware of his actual birthplace. However, by consulting Parish Rate Books, early enclosure maps and ordnance survey maps, it can be discovered that Sir Joseph was born on 28th March 1819 in Clay Hill, Enfield. Further detective work would suggest that the house of his birth was Hill Lodge, which once stood just below the Fallow Buck public house. Hill Lodge was demolished in 1965.

Sir Joseph William Bazalgette (1819–91).

After an education at private schools, Bazalgette became a pupil of Sir John Benjamin MacNeill in 1836, the same year in which he joined the Institution of Civil Engineers. MacNeill had begun his career under the famous civil engineer Thomas Telford and became one of his deputies. Telford clearly had a high regard for MacNeill as, after his death in 1834, it was discovered that he had remembered him in his will.

Bazalgette's early experience of work was on a drainage and reclamation project in the north of Ireland and as his career progressed he was given a range of other civil engineering tasks to complete. His career blossomed, as a pupil of MacNeill, and eventually the experience that he gained helped shape him into a highly respected civil engineer. By 1842, Bazalgette had set up on his own account in Westminster as a consulting engineer, being mainly engaged upon railway contracts.

In 1849, Bazalgette became a member of the Metropolitan Commission of Sewers. Appointments to this body were government nominations, which would suggest that his career had considerably progressed in a very short time. The new organisation had been set up in the previous year to replace

Hill Lodge, Clay Hill, Enfield, the birthplace of Joseph Bazalgette.

the eight different district bodies that had been responsible for the drainage of London. Previously, little thought had been given to creating a method of uniform drainage for the capital and as a consequence the system was extremely inefficient. Because there had not been one coordinating body charged with design and development, there were considerable differences of size, shape and fall of the sewers at district boundaries, with larger sewers discharging into smaller ones and egg-shaped sewers with narrow parts uppermost coupled to similar sewers built in reverse.

Up until 1815, it was illegal to discharge sewage or other noxious material into the sewers as it had been deemed they should only be used to carry surface water. At the time, cesspools were considered the only appropriate place for depositing household sewage. However, with London's increasing population it would seem that cesspools were unable to cope with the waste; and either the law was deliberately relaxed or a blind eye was turned by the authorities. It was not until 1847 that an Act was passed which made it compulsory to drain household waste into London's sewers. Within six years, thirty-thousand cesspools had been abolished as the effluent was directed into a modified, but not redesigned, sewage system. Now surface water and raw sewage combined to be discharged directly into the River Thames. In many instances, the outflows of sewage were close to where the private water companies took their supplies.

Between 1848 and 1855 the Metropolitan Commission of Sewers had been reconstituted six times with successive new appointments. This made it almost impossible for the body to implement any worthwhile schemes of sufficient magnitude that would have alleviated the growing problem of a highly polluted Thames. The water companies which took their supplies from the Thames came under increasing pressure as public concern for health intensified. In London, the total deaths from cholera alone, in 1854, amounted to almost twenty thousand. Although Bazalgette was appointed Chief Engineer of the fifth and sixth commissions, his early plans for solving London's drainage problems were frustrated.

In 1856, under an Act of Parliament, the Metropolitan Board of Works was set up. This body was the first attempt to bring about a system of local self-government by dividing the capital into thirty-nine districts. The City of London and the largest parishes, like Lambeth and Marylebone, formed separate districts, while the smaller parishes were amalgamated into further individual districts of manageable size. Bazalgette was appointed Chief Engineer to this new body and instructed to prepare plans for the drainage of London. This he duly did and the scheme was approved by the Board. However, Her Majesty's First Commissioner of Works had the power of veto and Bazalgette's plan was delayed. After much complicated negotiation and discussion, his recommendations were eventually adopted and work began on the scheme in 1859.

It is possible that Members of Parliament may have put pressure on those delaying progress, as, in the summer of 1858, when temperatures in the metropolis exceeded ninety degrees Fahrenheit, the stench from the polluted Thames became unbearable. Conditions became so bad that the windows of the Houses of Parliament had to be covered with curtains soaked in chloride of lime to try to overcome the dreadful smell. The episode became known as the 'Great Stink'.

Abbey Mills' east chimney; note the Second World War barrage balloon, top left.

Joseph Bazalgette's Abbey Mills Pumping Station, completed in the late 1860s.

Bazalgette's grand plan was to construct a sewage system which, as far as possible, would rely upon gravity and surface water to ensure the effluent was kept flowing. He also planned to divert the waste away from outlets which fed directly into the Thames near the centre of London, to a place some fourteen miles below London Bridge. He had calculated that, by discharging effluent at this distance, there would be little chance of the sewage being brought back to the metropolis by the turning tide.

As Bazalgette discovered, for the reasons outlined above, it was not possible to use the existing sewer network. He therefore arranged for new sewers to be constructed which consisted of 1,300 miles of brick-lined tunnels. On the north and south sides of the Thames, separate outfall sewers were built at a higher level so that they could carry the effluent to the discharge points further down the river. To raise the sludge from the lower-level gravity sewers, which were between thirty to forty feet below the new outfall sewers, powerful steam pumps were used. On the south side, the Southern Outfall Sewer was serviced by the pumping station at Crossness, while the Northern Outfall Sewer was supplied by the Abbey Mills pumping station, situated on the Essex side of the River Lea.

Presumably, in an effort to disguise the unromantic occupation of the Abbey Mills pumping station from local residents, the building was constructed on a grand scale. In its creation several different coloured bricks were used, the effect being so flamboyant that the building was nicknamed the 'Cathedral of Sewage'. Sometimes, locals called the building the 'mosque on the marsh', probably due to its iconic pair of tall, Moorish-style chimneys. The design was the work of Bazalgette and Edmund Cooper, the building being shaped in the form of a crucifix. Each of the four sections housed twin coal-fired beam engines and the exhaust from these was carried away by the two highly ornate chimneys. Although the Abbey Mills pumping station still stands, the two chimneys were demolished during the Second World War, leaving only their stone bases. It was thought that they acted as landmarks for the enemy, allowing bombers to pin-point targets when attacking the London Docks.

Before his death in March 1891, Bazalgette would have been rightly entitled to look back on a remarkable career with pride. He had restored, with an army of workmen, craftsmen and skilled engineers, the public health of London's citizens by eradicating water-borne diseases. However, he had not been able to foresee London's rapid population growth, from around 5.5 million at the time of his death to an estimated 8.6 million in 2017. Bazalgette could never have imagined, not even in his wildest dreams, how London and other major cities would have expanded upwards and horizontally, covering the ground with glass, concrete and steel. Nor could he have foreseen the coming of motor vehicles and the consequent network of tar-macadam roads, garages and freestanding paved areas.

Over the years, in the rush to build to satisfy population growth and evolving transport technology, planners and developers have ignored nature by building on flood plains and shrouding the land with concrete, which has led to major environmental problems in our towns and cities. When it rains, much of the water is now unable to penetrate the soil and make its way naturally into the underground storage aquifers; the surface drains are unable to cope with the excess and so they overflow. When this happens, surrounding buildings become threatened by flooding and sewage surcharges can occur within homes. At times of very heavy downpours, the excess runoff finds its way into the rivers which rise and breach their banks discharging, as nature intended, onto the flood plain where houses have often been built with little consideration for the consequences when a storm occurs. In London, when Bazalgette designed his sewage system, surface water from the drains was used to flush the effluent, under gravity, to sewage works for processing or to safe discharge points. As might be imagined, at times of heavy rain, the drains and sewers are unable to cope and severe surcharges occur which cause the authorities to discharge raw sewage into the River Thames as an emergency measure. Such deliberate polluting of the river occurs several times a year. It is a sad reflection on our planners and developers that we have come to this kind of solution when, over 150 years ago, Bazalgette designed a sewage system to keep the River Thames clean and free from disease.

Finding satisfactory solutions to these very severe problems is not a particularly easy or cheap undertaking. Fortunately, Thames Water has been able to come up with designs for two innovative, multi-million-pound schemes that will ensure the citizens of the capital will remain healthy for years to come. The first of these schemes is the Lee Tunnel project, begun in February 2012 and completed in January 2016. This is a 4.3 mile (6.9 km) tunnel that passes below ground, adjacent to Bazalgette's northern outfall sewer, and runs from Abbey Mills Pumping Station in Stratford to the Beckton Sewage Treatment Works at Newham. The tunnel has a diameter of 24 feet (7.2 m), the width of three London busses, and is constructed in such a way as to prevent effluent leakage and stop ground-water ingress. To achieve these high levels of protection, both tunnel and shafts consist of two fibre-reinforced concrete-lined segments: the primary being 350 mm thick and the secondary 300 mm thick. As it might be imagined, the Lee Tunnel has been designed to last and of course comes at a considerable cost.

Lee Tunnel under construction.

Lee Tunnel boring machine.

Lee Tunnel pump.

Lowering boring-machine section into Lee Tunnel.

The machinery to bore the tunnel was custom built in Germany and shipped to Tilbury across the North Sea by barge. From there it was transported to the drill site on large lorries taking in the order of sixty deliveries to complete the transfer. When fully assembled, the machine known as Busy Lizzie measures 120-metres long with a diameter of 8.88 metres. Once in operation, Busy Lizzie mixes one-hundred tons of excavated chalk with water for every metre of tunnel, completing seventeen metres of tunnelling in one day. The chalk slurry is removed to the surface by conveyor and piped to the Beckton end of the tunnel for processing before being shipped to its final destination by barge.

During heavy rainfall, excess surface water and raw sewage can now be stored in the completed tunnel rather than having to be discharged into the Channelsea River and then via the River Lea into the Thames. When the downpour subsides the effluent is pumped from the tunnel to the Beckton Sewage Treatment Works for processing. Here the solids are separated from the water and dried. The solid residue is compressed into cake which is burned in an on-site furnace which boils water to produce steam. The steam is fed, under pressure, to drive a turbine which is connected to a generator that produces electricity. This is an extremely environmentally friendly system as this energy is used to run the treatment works. Any excess power is sold to the NATIONAL GRID. The water that remains after separation is processed and discharged back into the river system and some will eventually receive further processing and arrive back in the household tap as drinking water. This is a recycling story at its very best, even to the degree that when Londoners flush their toilets they could be helping to keep the lights on!

Dumping sludge at sea in the 1950s.

Poo cake after the drying process; this will now be used to power electricity generators.

Making electricity from poo cake: the Crossness sludge-powered generator.

Beckton fat-fired power plant. Fat collected from sewers and drains is used to keep our lights on.

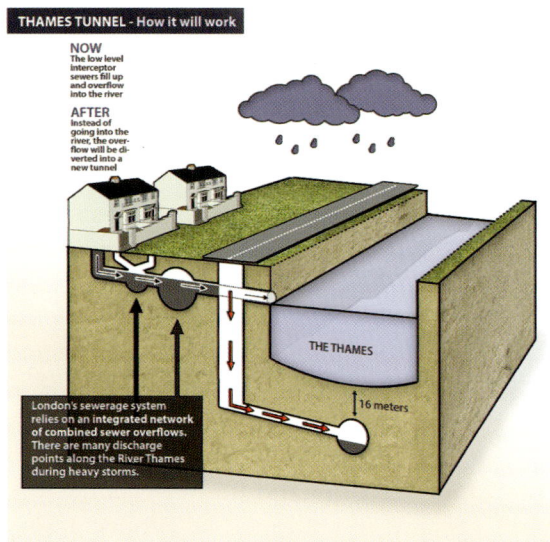

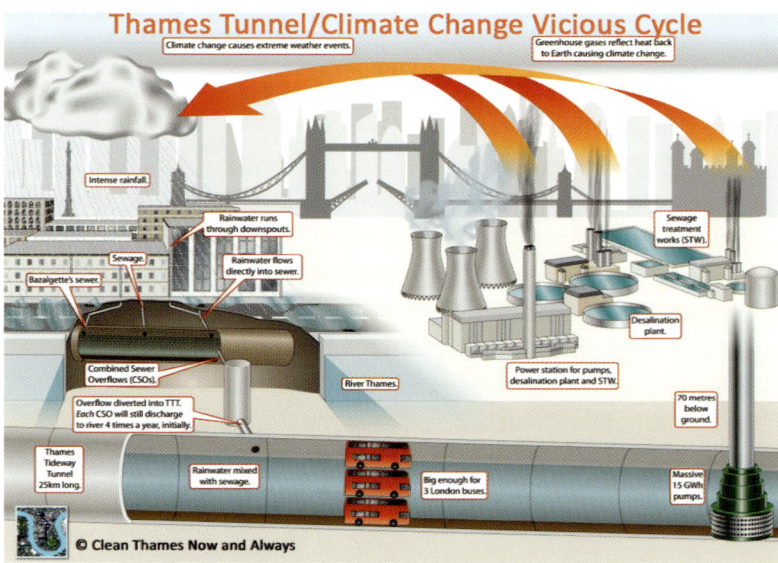

Diagram of how the Tideway Tunnel works. *Diagram illustrating the Thames Tunnel vicious cycle.*

The second Thames Water project is much bigger: the multi-billion-pound Thames Tideway Tunnel scheme. Starting in 2016, it is estimated that construction will finish in 2023, thus taking around seven years to complete. The planning, consultation and tendering processes took many years to reach fruition before an agreed route for the tunnel could be finalised. A sixteen-mile tunnel (25 km) will run mostly under the tidal section of the River Thames taking effluent from Acton in West London, joined by other combined sewer overflows (CSOs) along the way, to the Abbey Mills Pumping Station at Stratford. Here the effluent, composed of rainwater and raw sewage that would previously have been discharged into the Thames, will now be transferred to the Lee Tunnel for onward transportation to the Beckton Sewage Treatment Works at Newham. The Thames Tideway Tunnel dimensions and construction methods will mirror the same exacting standards as those of the Lee Tunnel project. When the whole project is completed, in approximately five years' time, it will future-proof London with a system of sewage capture, storage and environmentally friendly processing.

REFERENCES

Author unknown, *Riverside Sewage Treatment Works – Sewage Works Upgrades*, Thames Water, 2012

Author unknown, *Beckton Sewage Treatment Works – Sewage Works Upgrades*, Thames Water, 2013

Author unknown, *Lee Tunnel – London Tideway Improvements*, Thames Water, 2013

Author unknown, 'Green light for £4.2bn London "super sewer"', Thames Tideway Forum, 2015

Bazalgette, William Joseph, 'On the Main Drainage of London and the Interception of the Sewage from the River Thames', *Minutes of Proceedings of the Institution of Civil Engineers*, 14th March 1865, Volume 24 (1864–5)

Bazalgette, William Joseph, 'Address of Sir J.W. Bazalgette, President', *Minutes of Proceedings of the Institution of Civil Engineers*, 8th January 1884

Lee, Sidney (ed.), *Dictionary of National Biography*, Volume XXIL, Smith Elder & Co., London, 1909

Lewis, Jim, *London's Lea Valley: Britain's Best Kept Secret*, Phillimore & Company Ltd, Chichester, 1999

Lewis, Jim, *East Ham & West Ham Past*, Historical Publications Ltd, London, 2004

WAS THE LEA VALLEY ALWAYS AN INDUSTRIAL REGION?

Most people would probably believe that the Lea Valley region is famous only for its plethora of world-changing industrial and scientific firsts, but that would be a false assumption. The truth is the Lea Valley has a long and proud association with both horticulture and agriculture, and this actively encouraged a range of supporting industries and service providers to develop and set up in the region.

Referring to the Domesday Book, it can be seen that there was considerable farming and smallholding activity in the area as our distant ancestors began to learn the skills and techniques of planting and husbandry which helped to support and sustain a growing population.

It is hard to imagine that an area like today's Newham, the borough that hosted the 2012 Olympics and which currently appears to be developing by the day, as evidenced by its rapidly changing skyline, was growing potatoes on a commercial scale from the early eighteenth century. To help with the harvesting of the crop, Irish labour had to be imported to lift and collect the tubers which were becoming an important part of the daily diet of people in Britain.

Growing vegetables was fairly common in east London at the turn of the twentieth century and areas of cultivation seemed to crop up (pardon the pun) in the most unlikely places. During the nineteenth century, the fundamentals of agriculture were taught to the 200 boys of St Edward's Roman Catholic Reformatory School on the fourteen acres of land attached to the school building – Green Street House, East Ham. This agricultural land eventually became the home of West Ham United Football Club and remained so until the Hammers transferred to their new stadium on the Queen Elizabeth Olympic Park.

A Quaker, Dr John Fothergill (1712–80), purchased Upton House in West Ham (now Newham) in 1762 and set up home and also his doctor's practice. Here, in the grounds of his house (now West Ham Park) he established a botanical garden which, at the time, rivalled the Royal Botanical Gardens at Kew. Apart from growing exotic fruit and plants in hot houses that he had built, he

Dr John Fothergill (1712–80).

Corporation of London board, West Ham Park.

Rock Garden, West Ham Park, created in a style that Fothergill might have used.

planted shrubs and trees from around the world that could be used as a basis for new medicines and also food. Astonishingly, his plant collection amounted to over 3,400 different species. Sir Joseph Banks, the eminent botanist who had sailed with Captain James Cook on his first voyage of discovery (1768–71), was so impressed with Fothergill's work and ingenuity that he wrote:

> At an expense seldom undertaken by an individual and with an ardour that was visible in the whole of his conduct, he procured from all parts of the world a great number of the rarest plants, and protected them in the amplest buildings which this or any other country has seen.

Engraving of Upton House, later, Ham House, the site of Dr John Fothergill's botanical gardens.

The Lord Mayor visiting West Ham Park, September 1885.

The former Superintendent of West Ham Park, David Jones CBE, standing by the stone cairn close to the site where Upton House once stood.

Joachim Conrad Loddiges (1738–1826).

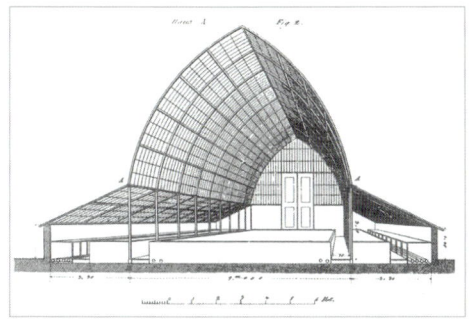

Loddiges' hothouse, Hackney, claimed to be the largest in the world at the time.

In the 1740s, Johann Busch (anglicised to John Busch or John Bush) arrived in Hackney from Germany and became a supplier of unusual plants to several botanical gardens, in particular those of Princess Augusta, daughter-in-law of George II. Augusta's plant collection was eventually to form the basis for the Royal Botanic Gardens at Kew. Busch's work appears to have got him noticed and he was invited to Russia by Catherine the Great and commissioned to lay out gardens in the 'English style'.

Not long after Busch, in the early 1760s, another German, Joachim Conrad Loddiges, a gardener, settled in Hackney. After working to landscape the grounds of Dr (later Sir) John Baptist Silvester, he set up in business as an importer of rare seeds and plants. Many of these he cultivated in large steam-heated hot houses that he had built on a site in Mare Street, Hackney approximately where the Town Hall stands today. When hardy, many of these plants were exported overseas in special packaging which Loddiges had designed.

In the early nineteenth century, plants from the Loddiges Nursery were sent to Madeira where they helped to establish a successful tea plantation on the island. As the popularity of the Loddiges Nursery grew, plants were supplied to the royal parks and also to the great estates of Woburn Abbey and Chatsworth. When the building used for the Great Exhibition of 1851 was relocated (in 1854) from Hyde Park to the site at Crystal Palace, Sydenham, a massive Mauritius fan palm weighing around fifteen tons left the Loddiges Nursery to adorn the building, making its way by road to its new destination pulled by thirty-two horses.

As the nineteenth century progressed, the march of industry continued, fuelled by the expansion of the railways. Pressure across east London mounted for sites to build new factories and also houses for the rapidly growing population that had been attracted to the area by the promise of work. This made smallholding and nursery land an obvious target for the developers. As the factories took hold in the lower Lea Valley, their presence caused increased levels of atmospheric pollution and severe poisoning of water courses. As might be imagined, this made the area less favourable to growers, who were eventually faced with either closing down or moving out. The attraction of cheaper land in the cleaner environment of the upper Lea Valley made districts like Enfield, Waltham Cross, Waltham Abbey, Cheshunt and Nazeing ideal places for established and new growers to put down their roots. There was also the added attraction that the London markets were only a horse and cart journey away.

From the beginning of the nineteenth century, a number of small nurseries were set up in the region of the upper Lea Valley. According to the late Peter Rooke, whose grandfather George had run nurseries in the Lea Valley, these were little more than open patches of ground where vegetables and bedding plants were grown out of doors. By the end of the century, with pressure on the lower Lea Valley growers to move due to the clamour of industry for building land, it had

A giant Mauritius fan palm on its way to Crystal Palace, Sydenham, south London, pulled by thirty-two horses, c.1854.

become clear that the upper Lea Valley offered the best opportunities for those who wished to start again.

As the upper Lea Valley growers' businesses had no doubt been drawn to this part of the region by its fertile loams and abundant water supplies, the area must have looked extremely attractive to those who were being forced out of areas of London polluted by industry. With the lower region's growers moving into the area alongside their upper Lea Valley counterparts, it would have probably seemed obvious to these early pioneers that, as their numbers increased, they could collectively become a dominant force in Britain's horticultural industry. However, for this to happen, the growers would need to come together as an organised group.

On Tuesday 31st October 1911, a meeting of representatives from the nurserymen and growers of the Lea Valley was convened at the Imperial Club, Waltham Cross and a resolution was moved by the Chairman and passed by those attending: "That an Association be formed which shall be called 'The Lea Valley and District Nurserymen's and Growers' Association'". The passing of a second resolution set the annual membership subscription at ten shillings and six pence (half a guinea) that would translate to 52.5 pence in today's money! Unfortunately, the minutes of the meeting do not list all those in attendance, so we are unable to compare the number of original members with that of today's Lea Valley Growers' Association (as the organisation has become known).

Michael Rochford (back row, second right) the founder of a Lea Valley dynasty of growers with his garden staff.

It is clear from the minutes that the nurserymen and growers were facing a number of quite serious challenges which they had reasonably concluded would only stand a chance of being resolved if they came together as a bona fide organisation. By forming an association, the growers felt that they would have a better opportunity to solve their problems collectively as certain government agencies would only recognise those organisations that were properly constituted.

The minutes of the first meeting reflect a number of outstanding concerns that the growers wanted to be addressed:

(a) The study of the various insect and fungoid pests and diseases which were yearly increasing

(b) The question of rating which had just been discussed at the meetings of the Cheshunt District Council

(c) Lectures

(d) Difficulties with the railway companies, market salesmen etc.

While later minutes show that not all the growers' problems could be resolved by coming together as an association, they do demonstrate that the organisation rapidly grew in stature and was acting in a professional and democratic way on behalf of its members. As early as January 1912, the organisation had been incorporated as a limited liability company and a programme of lectures arranged on subjects appropriate to the members' requests. Later that year, correspondence began with the Great Eastern Railway (GER) over its inadequate provision of goods sidings at local stations and also that company's inability to accept manure at Enfield Lock Station. Deals were negotiated with insurers and a ten percent discount on premiums for employer's liability insurance was secured for members with the Legal Insurance Company.

Rothamsted Research Centre, Harpenden.

By November, after listening to advice from Dr Russell of the agricultural Rothamsted Experimental Station, Harpenden, a special sub-committee of growers was formed. Soon the sub-committee was recommending that members should seriously consider a highly ambitious scheme to set up a local experimental station to carry out scientific work for the organisation. Should the recommendation be approved, costs for setting up and running the station would be shared partially by funds raised by the growers. The Association would also approach the Board of Agriculture and the County Councils of Essex, Hertfordshire and Middlesex for grants to make up the outstanding costs for the Station's continued upkeep.

In the following year, representation was made by the Council of the Association to the Board of Trade over the practice undertaken by some wholesalers and retailers of removing labels from tomatoes, cucumbers and grapes grown abroad so they could be passed off as English produce. Presumably, this devious removal of labelling was done so that the perpetrators could command a higher price for their produce. While it would be naive to believe that the Council could win every battle taken up on behalf of its members, the minutes show that its representatives went about their duties with a great deal of dedication and energy. They were certainly not afraid to tackle the major service providers over cost increases and there was no hesitation in confronting government departments on issues such as tax and legislation that affected the horticultural industry. The Lea Valley and District Nurserymen's and Growers' Association had begun the task of supporting its members with great enthusiasm, getting this newly formed body off to a flying start.

We have seen in this chapter that the Lea Valley region has provided a home not just for manufacturing industry, but for horticulture too. However, in a highly competitive and price-conscious world, will this balance of horticulture and industry remain for the foreseeable future? Answering this question is probably best left to those of a certain age who worked locally during the post-war era. As one of those people, the author will share his views, and also those of his colleagues as they have emerged during discussions over the past few years. In a nutshell, the rise and fall of Lea Valley manufacturing and the growth of horticulture during the post-war period can be looked upon as a microcosm of how Western industrial economy has performed generally. Those other countries that are currently developing their industries and those economies that mature later will no doubt pass through a similar phase. Only time will tell if these unscientific forecasts emerge as either true or false!

REFERENCES

Currie, C.R.J., (ed.), *The Victorian History of the County of Middlesex*, Vol. 10: Hackney Parish, University of London Institute of Historical Research, Oxford University Press, 1995

Lee, Sidney, (ed.), *Dictionary of National Biography*, Volume XXIL, Smith Elder & Co., London, 1909

Lewis, Jim, *London's Lea Valley: Britain's Best Kept Secret*, Phillimore & Company Ltd, Chichester, 1999

Lewis, Jim, *East Ham & West Ham Past*, Historical Publications Ltd, London, 2004

Lewis, Jim, *London's Lea Valley, a Century of Growing: The History of the Lea Valley Growers' Association*, Libri Publishing Ltd, Oxford, 2011

Solomon, David, *Loddiges of Hackney*, Hackney Society, 1995

CAN INDUSTRY AND HORTICULTURE SURVIVE AND CAN TECHNOLOGY PROVIDE ANSWERS?

Those people of a certain age, the author included, who worked in the period after the Second World War witnessed the significant decline in the Lea Valley region's manufacturing industries with the loss of skilled jobs. The making of things has been replaced by the growth of warehousing, retail parks, tourist attractions and service providers.

At first glance, it might be surmised that the horticultural industry suffered too, particularly if the demolition of the Rochford family nurseries at Turnford for house building is remembered. Nevertheless, it has not all been bad news. Today the organisation that brought the individual growers together, and became the Lea Valley Growers' Association, has collectively turned into a thriving horticultural industry supplying our shops and supermarkets with tomatoes, cucumbers, aubergines, peppers and leaf products. The reasons why they have been able to do this are twofold. One, the takeover of old and failing nurseries by second- and third-generation Italian and Sicilian families who now own ninety percent of the horticultural businesses; and two, the increasing and ongoing introduction of new technologies. Plants under glass are currently being grown in eco-friendly ways and the growers are continually striving to make their industry carbon neutral.

The following two horticultural examples, one small and one large, will give the reader an idea of how science and technology is being adopted to help provide us with healthy food while at the same time protecting and supporting our fragile environment.

Guy & Wright Limited

In the early years of the twenty-first century, Guy & Wright Limited, a small third-generation family tomato-growing business in Hertfordshire, owned by John and Caroline Jones, formed a partnership with Hennock Industries Ltd and New EnCo to develop a micro-turbine plant that would be powered from organic waste. To test the viability of the system, a working commercial plant was installed, part funded by the Department of Trade and Industry (DTI) under the Technology Programme. After successful trials, the plant was scaled up in size and John and Caroline took the bold decision to stop using fossil fuels as their main power source, retaining them as an alternative for use only when the economics dictated – which, to date, they have not had to do.

Installation of storage tanks for the anaerobic digestion plant (Guy & Wright).

Cucumber waste waiting to be fed into digester chamber at Guy & Wright.

Orange and lemon waste from markets used to fuel glasshouses (Guy & Wright).

Caroline and John Jones, owners of the family business Guy & Wright Nurseries, established in 1928.

Organic waste being delivered to Guy & Wright Nurseries.

Internal view of underground cell where organic waste is digested (Guy & Wright).

The organic waste reception at Guy & Wright.

To expand the system, known as anaerobic digestion (oxygen-free digestion), an enormous below-ground six-cell airtight bunker was constructed. Each cell is capable of holding 400 tonnes of organic waste. In the process of organic matter decomposition, a biogas is produced with high methane content which is compressed to directly drive micro-turbines. This system of electricity production is more efficient and cleaner than the traditional methods of making electricity from natural or town gas. Using this older technology, gas is burned to heat water to produce steam which in turn drives turbines that are coupled to the electricity generators. With the new system, the gas-burning stage of the process is completely eliminated.

The anaerobic digester in use by Guy & Wright is an exceedingly greedy animal, requiring fifty tonnes of organic material per day to satisfy its hunger. It will be appreciated that it would be impossible for the nursery to generate such a large daily amount of organic waste from its own operation, so arrangements have been made with London's Spitalfields Market, Bedfordshire Growers, who pre-pack onions, and the banana importers J.P. Fresh of Dartford. These companies, and others, are all extremely grateful to be able to donate their organic waste to Guy & Wright rather than send it to landfill, for which they would have to pay a levy. As an environmental initiative, it is an all-round win–win situation.

Waste material coming to the nursery is emptied by the delivery vehicles into a receiving pit and from there it slowly moves to a macerator where it is crushed and broken down into smaller pieces. The resultant 'vegetable soup' is pumped into a holding tank before being distributed to the six underground digesting cells where, over a three-week period, bacteria break

Straw boiler installed to heat the nursery and reduce the use of expensive gas (Guy & Wright).

Emptying organic vegetable waste from markets into the digester chamber (Guy & Wright).

Tomatoes are grown at the nursery nine months of the year (Guy & Wright).

down the vegetable material. This process is closely controlled by computer probes and only after testing and analysis does the required amount of material get fed to each of the cells. Waste residue from the cells is fed to three established reed beds, allowing the system to supply gas continuously without the need to shut down for cleaning. The biogas generated by the plant is fed into a giant inflatable bag which can then supply the five micro-turbines that produce the electricity and also a small boiler which is used to power the three miles of underground heating below the digestion cells. The CO_2 produced by the process is so clean, being mainly devoid of sulphur, that it can be fed directly into the glasshouse growing area to be absorbed by the plants during the day without any form of treatment. This arrangement has considerably improved crop yield.

The installation of the plant, which began in 2003, took three years to complete with much of the construction work done by the Joneses and local specialist contractors. As expected, with a plant so unique, initial teething problems occurred and these had to be resolved alongside the Jones family running their business. In what appears to be an understatement, Caroline Jones says that the plant installation "has been by no means easy" but now she concedes, with, I suspect, some relief, that "we have achieved our goal".

Dr Andy Marchant of Hennock International Limited, the designer of the anaerobic digestion system at the nursery, has claimed that "John and Caroline Jones were the first growers in the world to install micro-turbines with high-rate CO_2 enrichment on a commercial nursery, and have been host to countless other growers keen to see what they were doing".

When the plant became fully operational, the system produced more than enough electricity to satisfy the running of the nursery and the excess power was sold to Green Energy, a company that was established specifically to obtain and distribute electricity from renewable sources. This arrangement allows the Joneses to forget about the need for customer administration, with its inherent problems of metering and billing, and to get on with the daily task of running their business. For all their hard work, considerable financial expenditure and futuristic risk-taking, John and Caroline Jones rightly deserved the prestigious Grower of the Year Awards 2009 – Business Initiative of the Year commendation. In February 2016, Guy & Wright also won the Energy Now Expo 2016 Best Anaerobic Digestion and Biogas Scheme category award for their continuing commitment to producing commercial tomatoes in an eco-friendly way.

On a recent visit to Guy & Wright in January 2017, the author learned that a significant improvement to the efficiency of electricity production had been made which was achieved by adding a simple and cost-effective facility. Waste residue from the digester is now harvested in a lagoon and the last remnants of biogas are extracted and saved in an inflatable bag. Caroline Wright explained that squeezing out this extra gas has allowed them to reduce the tonnage of vegetable waste arriving daily and has increased the export of electricity to the National Grid to one megawatt. Caroline also explained that without making such efficiencies, particularly in the current economic climate, the business could not survive if they had to rely on traditional mains gas supplies.

Tomsworld Ltd

The second establishment that I visited, Tomsworld Ltd, Nazeing, Essex, part of the Glinwell PLC group of nurseries, began the first phase of building in 2012 on a former nursery site. A second phase of building has now been

Work begins on the Tomsworld nursery site at Nazeing in 2012.

completed and a third phase has already been planned and the land levelled for building.

The current nursery boasts twenty acres of glass and when the third phase is completed another eight acres of glass will be added. For the Lea Valley, this scale of glasshouse building is truly staggering. When viewing these glasshouses from the inside, the lines of what are termed "raised gutters", which are designed for growing plants, vanish into the distance in all directions surrounding the viewer.

Glinwell PLC has invested heavily in a series of these hi-tech glasshouses where aluminium has replaced the traditional wood in the construction and they have also installed the latest state-of-the-art combined heat and power systems (CHP) to replace the old labour-intensive and inefficient boiler heating systems. Here glasshouses with floor spaces that are measured in acres rather than metres are growing tomatoes as their main crop. Gas has taken over as the preferred fuel from the old solid fuel and oil burning systems that were previously employed to heat the glasshouses. The new gas-fired boilers are fitted with condensing units which capture the exhaust gasses, which are then cooled and cleaned before being expelled to the atmosphere outside mostly as hot air and water vapour. Flue gasses, mainly carbon dioxide (CO_2), extracted and cleaned by the condenser are pushed through a maze of pipework into the growing areas. This is naturally absorbed by the plants during the day providing a very efficient method of carbon capture. The grower has found that enriching the internal glasshouse atmosphere by up to 2.5 times the natural level can improve crop yields. Surplus electricity generated by the CHP system is sold to electricity supply companies and fed into the national grid.

Plants are grown in individual containers that are carried by raised gutters. Each plant is automatically fed a controlled amount of water and nutrients

Water filtration and monitoring machinery which ensures that plants are supplied with quality nutrients (Tomsworld).

The yellow box in the foreground contains bees, nature's pollinators (Tomsworld).

Site drainage ditches preserved to provide habitats for water voles, amphibians and grass snakes (Tomsworld).

Nutrient storage and distribution equipment (Tomsworld).

under a computer-controlled and pressure-regulated drip-feed system. No matter where the individual plants are within the glasshouse, the computer is programmed to ensure that each one receives an exact amount of nutrients. The containers, which are filled with a rock-wool fibre mixture rather than earth or compost, are placed on the lines of raised gutters. These are effectively exceptionally long shelves, suspended above the ground on a system of wire braces connected to the framework of the glasshouse. As early as the eighteenth century nurserymen began to understand how plants absorbed their essential nutrients for growth through water. Soil only acts as a sort of holding reservoir for the nutrients and so is not essential to a plant's growth. Growing by this system is called hydroponics and has many advantages for the nurseryman. Having the control of the nutrient levels also reduces cost, improves crop yield and releases less pollution into the environment. Not having soil allows better control of pests and diseases as these can lie dormant in the ground to spring a nasty surprise on the grower later.

Water is effectively recycled as it is kept within the system, thereby reducing the amount that needs to be taken from the on-site deep borehole. Rainwater is harvested from the acres of glasshouse roofs and stored in large tanks, then, after filtering and sterilising, it is pumped into clean-water storage tanks before being fed to the growing areas. Also, outside rainwater and surface water are collected in a large over-ground reservoir stocked with fish that help with the cleaning process. This water, after filtering and sterilising, is then available for use in the glasshouses. Running nurseries on this scale to produce home-grown food is an expensive business; and as we consumers increasingly demand low-priced produce, savings made by harvesting water have benefits for both consumer and environment.

Modern glasshouses that operate the CHP system employ thermal screens. These have around an eighty percent light transmission and are fitted approximately 3.5 metres above ground level and 1.5 metres below the roof to reduce heat loss. The screens are computer controlled, as are most of the growing operations, being pre-programmed by the grower to respond to the internal and external variations of temperature and light within the particular

The Tomsworld combined heat and power (CHP) plant employing gas condensing boilers that extract and recycle energy.

The turbine room where electricity is generated from methane gas produced from organic waste (Tomsworld).

Matthew Simon, manager of Tomsworld Ltd, Nazeing, standing in a recently planted glasshouse.

glasshouse. Every opportunity has been taken to ensure that all surfaces within the glasshouse – including floors, raised gutters and the framework supporting the glass – are light in colour to reflect every photon of available light. I was informed by Matthew Simon, the site manager, that the level of light within the glasshouses was actually brighter than that on the outside.

Anyone visiting the Tomsworld site could not fail to appreciate the efforts that have been made to encourage wildlife to visit the site. When the ground was cleared for building, oak trees were left standing and now support bird boxes. Natural watercourses and their wildlife habitats were preserved and now water voles, grass snakes and adders thrive.

Science now plays a big part in maintaining a healthy glasshouse crop and chemical usage for pest control is kept to an absolute minimum. The modern grower uses a system of Integrated Crop Management (ICM) where biological predators are used to control the pests. This not only saves money by reducing the need for expensive chemicals: it also provides a healthier product for the health-conscious consumer. Bees, nature's highly efficient pollinating insects, have been introduced into the glasshouses to ensure that the plants are fertilised as nature intended.

When entering one of these glasshouses for the first time it would be impossible not to be overwhelmed, both by the vastness of the place and the industrial scale of plant growing. The clean and clinical conditions of the growing area are not unlike those one might expect in a modern hospital. Another feature which the visitor would not fail to notice is the distinct lack of labour. Of course, people have to be used to hand string the tomato and other plants, and to take off excess leaves. When the growing seasons ends, the plants have to be removed and composted and the glasshouses cleaned and sterilised to make the area ready for the next crop, but otherwise very little labour is required.

Water tanks holding surplus rainwater that has been harvested from roofs and lagoons (Tomsworld).

Some nurseries use a system of grow-bags which are planted with specially grown seedlings cultivated by specialist producers. For example, plants like cucumbers grown in this manner can produce three quality crops in a season.

Naturally, all these improvements have come at a cost and growers have had to raise millions of pounds to cover their investment but it is the only way for the horticultural industry in Britain to survive when competing with their counterparts in hot countries that have the luxury of smaller heating and lighting bills.

Over the past decade there has been a decline in the number of Lea Valley horticultural growers but overall they are still supplying the UK markets with a third of their cucumber requirements. Also it should be remembered that in the 1970s the yield for tomatoes grown in the region was approximately 70 tons per acre. Now, with the new technology, it is typically 280 tons, with more varieties grown to satisfy changing consumer demands and taste.

It is not possible to predict the state of the Lea Valley horticultural industry over the next ten years, particularly in the current uncertain economic climate, but the core of growers who have made the commitment to provide us with salads produced in the most environmentally friendly way deserve the support of all the retailers and also us, the consumer.

For industries that supply a consumer market, there often comes a point when the requirements of that market change and for the growers this can provide a dilemma. Currently, environmentally aware consumers are making demands on retailers for more fresh produce that has been grown in Britain. Apart from home-grown food preserving the freshness and taste of the product, it also considerably reduces the high mileage the product has to travel. This helps to reduce the carbon footprint of food growing. I am constantly being told by growers that UK shops and supermarkets will take all the home-grown produce they can provide but, in many cases, the growers' hands are tied. The reasons for this are two-fold. Firstly, technology has just about reached the maximum crop a grower can achieve in a specific space and so there is an urgent need to extend the growing area. This means building new glasshouses – which is where problem number two occurs.

Ironically, growers who want to increase the size of their nurseries to cope with consumer demand for British-grown produce often fall foul of government and local authority planners who seem reluctant to sanction new build schemes quickly. To help overcome these problems, growers have resorted to snapping up just about every old nursery that comes onto the market in their quest to expand, as we have seen earlier with Tomsworld. Unfortunately these opportunities are few and far between so growers frustratingly have to sit and twiddle their thumbs, or take the expensive option of moving their business to another part of the country. While these moves and delays are taking place, overseas growers can easily fill the supply gap.

One alarming story that I heard recently illustrates the planning morass that regularly heightens frustration levels amongst growers. A local authority gave planning permission to build a fifty-bed care home on land designated in the local plan for horticultural use. The site in question had once been occupied by a nursery which a number of local growers had wanted to buy to expand their businesses.

Not too long ago we could only buy fresh fruit and vegetables that were in season but all that has changed dramatically. When we walk into our local shop or supermarket, we expect to be able to purchase all the different fruits and vegetables all the year round. Should we want to have fresh runner beans and garden peas and perhaps an exotic fruit with our Christmas dinner, then all we have to do is pop to the shop. Perhaps it is too much to ask our members of parliament and our local government officers to work together to draw up a long-term strategy for an in-house food industry with a view to reducing imports of produce that we can easily grow ourselves. If such a strategy *could* be agreed, it would have the added impact of cutting our food miles, but for the time being we will probably have to dream! However, we can all play our part in protecting the environment and helping our wildlife, without too much effort, by supporting our British growers and ensuring we eat healthily.

REFERENCES

Interview with Caroline Jones, Guy & Wright Ltd, Green Tye, Hertfordshire, 12th January 2017

Interview with Matthew Simon, Tomsworld Ltd, Pecks Hill, Nazeing, Essex, 12th January 2017

Lewis, Jim, *London's Lea Valley, a Century of Growing: The History of the Lea Valley Growers' Association from 1911 to 2011*, Faringdon, Oxfordshire: Libri Publishing, 2011

SELLING CARBON CAPTURE

The more I write and research the Lea Valley region, the more I come across stories that have been lying dormant for years and are begging to be told. The one that I am about to reveal is a prime example. I had thought that the Rochford name had disappeared from the Lea Valley when the Rochford nursery businesses at Turnford, Hertfordshire succumbed to the late post-war land-grab for housing, as new industries and warehousing marched northward up the Lea Valley. Fortunately, I was mistaken.

In 1949 the firm of Duncan Tucker, Lawrence Road, Tottenham, a prominent manufacturer and supplier of glasshouses since 1830, bought seven acres of land at Letty Green, Hertfordshire, turning the grounds into a demonstration site for his business. By 1959 Joseph Rochford & Sons had bought the business from Tucker's company and converted the site into a propagation facility that was needed to supply their thriving Turnford nurseries. In the mid-1970s the business at Letty Green changed from a propagation facility and moved into hardy nursery stock production, becoming a supplier to garden centres, local authorities and specialist landscape designers, eventually turning into one of the country's first wholesale cash-and-carry suppliers. Joseph Rochford took this change of direction in order to make use of a small nursery that was no longer needed as a supplier of young plants for the ever-decreasing Turnford nurseries. As the new business expanded, adjacent land was acquired.

By the mid-1980s, the Joseph Rochford & Sons business at Turnford had finally succumbed to the bulldozers of the housing developers and, in 1988, the business at Letty Green was taken over by Paul Rochford. Paul is the great-great-grandson of Michael Rochford, the man who founded this dynasty of nurserymen when he came to Britain in 1840 from Ireland, just escaping the

Entrance to Rochfords Nursery, Letty Green, Hertfordshire.

Paul Rochford standing beside the painting of his great-great-grandfather, Michael, the founder of the Rochford dynasty of nurserymen.

peak of the potato famine. The company now trades under the name Joseph Rochford Gardens Ltd, keeping the respected family name alive and well; so well in fact that an additional site was acquired, increasing the production area to its present fifty acres. Paul Rochford currently holds the position of Master of the Worshipful Company of Gardeners (2016–17) and is also the President of the Royal National Rose Society.

By 1996, under Paul's initiative, a new production facility was created to take advantage of modern water-conservation techniques. Both sites now boast a system that harvests water runoff from buildings, nursery roofs and lined drainage ditches. This water is filtered through reed beds and then stored in on-site lagoons. The water is regularly monitored and analysed for contaminants, which if found are filtered out, before being pumped back into the sub irrigation beds that supply the plants. There are two bore holes that help to top up the lagoons which can lose water in hot weather due to surface evaporation.

When the business changed to the production of hardy nursery stock, specialising in shrubs, trees, herbaceous plants and grasses, it was able to eliminate expensive heating plant and equipment almost immediately. The business grows its stock outside and in unheated glasshouses and poly-tunnels and has the privilege of not having to compete with hotter countries, unlike the present UK growers of our salads who have to ensure their glasshouses are kept at regulated temperatures due to our inclement weather. This is probably one of the reasons why, in such a relatively short space of time, Rochfords has become one of the largest wholesale nurseries in the UK: it ensures strict economic resource management which helps it to remain competitive.

By 2006, Paul Rochford had given some serious thought to the greening of his nursery and decided to replace the diesel mini tractors, used for moving heavy on-site products, with new electric tow trucks. In 2012, Rochfords installed solar panels which annually produce twenty-five percent of the electricity consumed on site. The solar arrays are photo voltaic which means that electricity can be generated when cloud covers the sun. At times when the light input is high, surplus electricity can be sold to the National Grid, helping to keep costs down. 2014 saw the installation of a pellet biomass boiler which has reduced red diesel usage by forty percent. These initiatives have seen a reduction in CO_2 emissions by around 42,500 kg per year.

Heavy plant brought in to construct a new lagoon for rainwater harvesting.

The company operates a policy of using bio-control methods when growing its plants and chemical usage is kept to an absolute minimum. Typically two different types of growing media are used on site: a one hundred percent peat-free growing media and a thirty percent peat-free growing media. These peat-free and peat-reduced mixes are used to nourish the particular plant species being grown. The company aims to maximise the natural defence mechanisms of a plant to help it combat pest and disease infections, and only uses pesticides as a last resort.

Nurseryman being kept busy potting shrubs.

Bedding plant preparation in one of the new poly-tunnels.

Part of one of the vast poly-tunnels that are required to satisfy the constant demand for plants, shrubs and trees.

In order to ensure that the two sites perform to high standards, Rochfords became a member of the British Ornamental Plant Producers (BOPP) scheme in 2004. BOPP is a certification scheme for growers, packers and manufacturers of growing media. Independent inspections are carried out by the United Kingdom Accreditation Service (UKAS), an accredited ISO Standards certification body.

The Rochford story at Letty Green appears to be one of those unique tales that can happen when, due to external economic pressures, a successful business adapts its traditional product base – in this case, from growing plants that require a controlled hothouse environment, to the growing and selling of trees, which have 350 million years' experience of carbon capture! With deforestation around the world and tree diseases like ash dieback, we really need nurseries like Rochfords to help us capture CO_2 if we want our planet to survive.

REFERENCES

Grant, Rebecca, Special Projects Manager, Joseph Rochford Gardens Ltd, Interviews and correspondence, February and March 2017

Lewis, Jim, *London's Lea Valley, a Century of Growing: The History of the Lea Valley Growers' Association from 1911 to 2011*, Libri Publishing, Faringdon, Oxfordshire, 2011

TOURISM – THE LEA VALLEY'S ALTERNATIVE ECONOMY

The Royal Gunpowder Mills at Waltham Abbey occupy some 190 acres towards the middle of the Lea Valley and were in continuous government use from 1787 to 1991. Before this, gunpowder had been produced on the site by the Walton family and, preceding that, by the Hudsons from the 1660s.

Until steam was introduced in the 1850s, water from the River Lea powered the machinery and charged the site's dual-level canal system. In the twentieth century, the development and manufacture of a range of chemical propellants and explosives, particularly cordite and RDX, were essential to the war effort. The latter explosive was used in the bouncing bomb dropped by 617 Squadron (the Dambusters) during the Second World War. The site, now decommissioned, contains over 300 structures, of which 21 are listed.

The experimental, development and production work carried out at Waltham Abbey has, over the centuries, provided employment and brought economic and social prosperity to the area. Moreover, this work wasn't exclusively military. The peaceful use of explosives in civil engineering has speeded the construction of roads, bridges, railways, canals and harbours and made the extraction of essential raw materials considerably easier in the mining and quarrying industries. Due to the quality of Waltham Abbey's products, which were recognised internationally, new export markets were opened up to Britain.

Government Row, former Royal Small Arms Factory workers' cottages.

With the closure of the Mills, economic hardship came to the locality. The question then had to be addressed: could any commercial alternatives be put in place? Solutions to such problems often depend upon a particular site's heritage and its accessibility to transport. If a stricken industry or place of manufacture has an interesting historic past, there is the possibility of developing a long-term strategy based on aspects of its heritage.

Napoleonic battle re-enactment at the Royal Gunpowder Mills, Waltham Abbey.

Waltham Abbey is truly a jewel in the tourism and heritage crown. The site was recognised by English Heritage as "the most important to the history of explosives manufacture in Europe" and the Royal Commission on the Historical Monuments of England (RCHME) has said that the Gunpowder Mills are "the most complex industrial site yet surveyed by the commission". There are also at least thirty-eight A4 pages of references to individual historical documents relating to the Waltham Abbey Mills located at the National Archive, Kew. As those who seriously study history will know, such a wealth of quality documented material is extremely rare.

As it had remained in government hands for over two hundred years, and due to the highly secret nature of the work undertaken there, the site took on the mantle of a forbidden city. External contractors for repair and maintenance were seldom used, the work being carried out internally by a highly skilled workforce that had to comply with the rigours of the Official Secrets Act. These were major factors in the conservation of the site which have clearly helped to guard the integrity of the artefacts.

When a particular source of power for the site's machinery became obsolete, or outlived its useful life, it was replaced with the latest technology. The abandoned machinery or technology was then discarded and allowed to fall slowly into a state of decay. This has provided a unique opportunity to trace, study and record the technological changes to the energy sources which drove the various pieces of processing plant, from water through steam to electricity. Apart from the obvious educational and academic benefits to be derived from studying such work, much can be learned from taking a hands-on approach to our past as it allows making comparisons with evolving present-day industrial and technological developments. By a careful analysis of the results it is often possible to learn important lessons which allow us to avoid repeating costly mistakes.

The northern end of the Gunpowder Mills site has been designated a Site of Special Scientific Interest (SSSI) by English Nature. Planted in the seventeenth century with alder trees to provide charcoal for the production of gunpowder, the wooded area once supported one of the largest heronries in Essex but unfortunately the herons have left to take up residence elsewhere in the Lee Valley Park. A scheme is currently in hand to help bring these birds back. However, the alder-tree seeds provide a favourite food source for the siskin (a small yellow–green bird not unlike the greenfinch) and the site now has some of the largest flocks of this bird to be seen in the UK. Protected by the dense vegetation, a variety of mammals liver there including deer, foxes, badgers and otters. In a deep pond called Newton's Pool, once used for testing under-water explosives, there are giant pike and other fish that have escaped the angler's hook.

After an extensive programme of decontamination costing £16 million, paid for by the Ministry of Defence, a charitable trust was formed to own, conserve and manage the Gunpowder Mills (north site) for the benefit of the public. A further role for the Trust was to formulate a strategy for the creation of a large visitor attraction. This was done by exploiting the site's heritage through the promotion of a plan which interpreted the various processes used in the manufacture of gunpowder. This initiative was boosted by a grant (in 1997) of £6.5 million from the Heritage Lottery Fund. However, the story of Waltham Abbey does not end with one visitor attraction; another arose out of man's destruction of nature.

On 28th July 1945, the Royal Gunpowder Mills closed and on 30th July reopened as an experimental wing of the government's Armament Research Department. This was a major change in status for Waltham Abbey as it moved from what was effectively a manufacturing facility to one that would become a world leader in experimental science, not just in explosives, but in advanced chemical engineering and rocket propellants. Much of this work was carried out on the facilities that had been established on the south side of Waltham Abbey, thus being known to the workforce as the south site (the original gunpowder mill complex was known as the north site). To achieve the new objectives, graduate engineers, scientists and chemists were recruited and soon became established as a highly regarded scientific development team. By October 1946, Waltham Abbey had become the Chemical Research and Development Department and would play a major role in designing and experimenting with a range of new materials that would, in the longer term, have a considerable spin-off both for the military and also for industrial and domestic markets.

The wartime rocket attacks on Britain and the development of fighter aircraft equipped with jet engines meant that methods of warfare and defence would never be the same again. Strategies would have to change dramatically. It was no coincidence that Waltham Abbey was now placed at the forefront of the emerging technological challenge. For most people this was an anxious period when the world held its breath as it entered a sensitive and critical phase in history – a phase that would become known as the Cold War, which in turn was to provoke an arms race between East and West.

As the Western world's relations with the East became increasingly strained, the emerging technology of the rocket, as a new form of defence and attack, was being developed by both camps. Interestingly, these developments were ably assisted by scientists who had previously worked within the German V2 missile programmes. These people had either been taken or had gone voluntarily to countries like Britain, the USSR and America.

The push for technological supremacy seemed to begin in earnest with the successful launch, in October 1957, of Sputnik by the Soviet Union, the first artificial satellite to orbit the Earth. This was followed a month later, again by the Soviets, when a live dog was carried into orbit. Naturally, much publicity was given to these events and almost four months later America was forced to respond with the launch of Explorer One, its first satellite. It is probably fair to say that what has become known as the Space Race had begun in earnest.

The Skylark rocket: its propellant was developed on what is now the Gunpowder Country Park, Waltham Abbey.

In the meantime, at government establishments like Farnborough in Hampshire, Westcott in Buckinghamshire and Waltham Abbey in Essex, things had not stood still. Although not as dramatic as the Soviet and American satellite launches, Waltham Abbey scientists, along with their sister establishments, had quietly developed the Skylark rocket that was first successfully launched from the rocket range at Woomera, Australia in February 1957. The Skylark had been designed to carry out a range of scientific experiments in the upper atmosphere to understand the significance of such phenomena as ultraviolet, X-ray, gamma and infrared radiation. A greater knowledge of these subjects helped pave the way for the exploration of space.

As technological requirements of the military changed, work on liquid rocket propellants transferred to the Rocket Propulsion Establishment (RPE) at Westcott and Waltham Abbey began work on new composite propellants and materials, which included those for safely blasting ejector seats from aircraft. In 1973, the establishments at Waltham Abbey and Westcott were merged. This was followed, in 1977, with further structural changes as Waltham Abbey became the Propellants, Explosives and Rocket Motor Establishment (PERME). This new set-up continued until 1984 when the government decided to privatise the Royal Ordnance factories and Waltham Abbey became divided between the old north site and the south site which then came under the control of Royal Ordnance plc.

Marshall steam engine driving overhead line shafting at Walthamstow Pumphouse Museum.

Although the government stopped funding the Skylark programme in 1977, believing that the American Space Shuttle would carry out future research, there was still a future for the project. Skylark was taken over by British Aerospace and later by Matra Marconi Space and became one of the most successful long-term rocket programmes in history, lasting almost fifty years. The final launch, its 441st, under the auspices of the European Space Agency, took place from Sweden in May 2005. Interestingly and somewhat ironically, the former site of the Skylark Rocket propellant development has become a new country park, the Gunpowder Park, administered by the Lee Valley Regional Park Authority. Strolling through this park on a pleasant summer's day, the visitor will often hear the thrilling song of the Skylark which has now

Markfield engine house and cafe, Tottenham. A good time to visit is when the Victorian beam engine is in steam.

Bazalgette's Abbey Mills Pumping Station, Stratford.

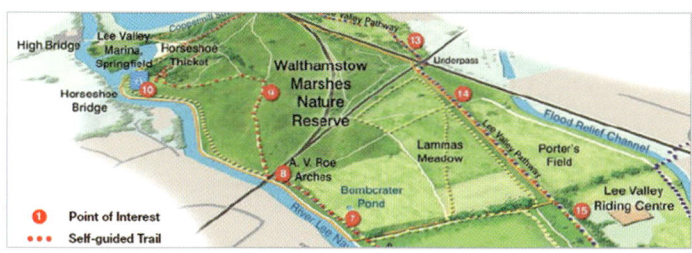

Map of Walthamstow Marshes.

been attracted to this new grassland habitat. Who could have imagined that this beautiful birdsong would ever be heard again in an area where Skylark rocket motors once roared? Meadow Pipit nest along with Skylark on the grassland in the summer months and the wildflowers amongst the grasses attract all kinds of insects with their nectar including butterflies. Look out for Common Blue, Holly Blue as well as Essex Small and Large Skipper.

At the northern end of the site a drainage channel, known as Black Ditch, flows and is home to water voles. On warm evenings, Daubenton's Bats, one of six species of bat on Gunpowder Park, can be seen hunting insects low over the water. A wet woodland of Osier Marsh has been developed on the Park with a boardwalk provided for the visitor. Here wintering Thrushes, Woodcock and large mixed flocks of Finches and Tits can be seen.

Cody Dock sensory garden, Lower Lea Valley.

Chadwell springs, north of Ware, source of the New River.

William Morris Gallery, Walthamstow.

Three Mills, Bromley by Bow, a good place to start to discover the Lower Lea Valley.

Walthamstow Wetlands, Coppermill Pumping Station, now a viewing platform.

White-water rafting near Waltham Abbey, a facility built for the 2012 Olympics.

Walthamstow Marshes nature reserve.

In this chapter we have seen how mankind's deliberate interference with nature has accidentally provided us with two priceless environmental assets, over two hundred years later. However, when we look towards tourism as opposed to manufacturing as the new revenue generator, it is appreciated by many agencies that this change of emphasis will not necessarily bring instant economic benefit to the region. Nevertheless, in the longer term it is hoped that the sensitive restoration of many historic industrial and other important sites within the Lea Valley will create a focus for heritage and promote the tourist industry. This in turn could attract other job-creating businesses to the area. After all, we have the model of Ironbridge Gorge and Coalbrookdale, where the unique industrial heritage of that region helped secure manufacturing investment, particularly from abroad, in the nearby town of Telford.

Another initiative within the Lea Valley region designed to attract tourism is the creation of new nature reserves and wetland habitats. Bodies such as the Lee Valley Regional Park Authority, the London Wildlife Trust, the Hertfordshire & Middlesex Wildlife Trust and others are heavily involved in creating and developing such programmes that will not only benefit wildlife but will also educate humans. There is the recent example of opening up the Queen Elizabeth Olympic Park to house other revenue-generating bodies. The Olympic Stadium is now the home of West Ham United Football Club, a major London museum has planned an extension on the site, a prominent university is to open a new campus and the BT media centre has taken up permanent residence. The Waltham Abbey White Water Rafting Centre, specifically built to host 2012 Olympic events, is now attracting visitors from across the UK. This, in turn, is helping the economy of the nearby town.

Having all these facilities within the Lea Valley makes the region an attractive place to live, work and play, thereby creating a ready workforce for future employers who are looking to set up new and innovative industries.

REFERENCES

Author unknown, *Where to Watch Wildlife in the Lee Valley*, Lee Valley Regional Park Authority, 2012

Lewis, Jim, *London's Lea Valley: Britain's Best Kept Secret*, Phillimore & Co. Ltd, 1999

Lewis, Jim, *London's Lea Valley: More Secrets Revealed*, Phillimore & Co. Ltd, 2001

Lewis, Jim, *From Gunpowder to Guns: The Story of Two Lea Valley Armouries*, Middlesex University Press, 2009

Patrick, Cath, Senior Conservation Officer, LVRPA, personal conversation, December 2016

SAVING THE PLANET – IS RUBBISH THE ANSWER?

No story about environmental issues, which are intrinsically linked to the public health of the nation, would be complete without trying to understand the problems that all of us have created, and continue to create, for our planet by the generation, transportation and disposal of household and industrial waste. Every year, in Britain alone, hundreds of millions of tonnes of waste are produced with large quantities of unprocessed material ending up in landfill sites, piling up environmental problems for the future. As we start our journey into the twenty-first century, nations around the world are beginning to wake up to the increasing dangers created by the melting of Artic ice sheets, rising sea levels and severe droughts, all brought about by the effects of global warming, caused by increasing emissions of methane (some from landfill sites), carbon dioxide (CO_2) and other polluting gasses such as CFCs (chlorofluorocarbons) which are destroying the protective ozone layer that surrounds our planet.

While the links between public health and waste disposal might not be immediately obvious, it is worth remembering that decaying and festering matter attracts vermin and can also lead to the spread of diseases. So it is important that our waste is disposed of in a controlled and efficient way that is sympathetic to the environment. This means recycling, rather than destroying, our precious materials and also making sure that toxic waste is not allowed to enter the food chain or water supplies.

Humankind has been increasing pollution levels since time immemorial but the process speeded up considerably with the start of the Industrial Revolution in the eighteenth century. The burning of large quantities of fossil fuels, particularly coal, to power our factories and power stations and the use of coal in the production of town gas to light and warm our homes have, over the years, allowed the release of millions of tonnes of carbon dioxide that had been stored naturally within fossil deposits. This harmful gas was trapped in the fossil materials for millions of years and our avaricious quest for more and more energy has secured its release into the atmosphere causing our planet to warm.

Over the last two centuries, there has been a rapid increase in deforestation in countries around the world to provide timber and also land for farming, in particular large-scale agricultural schemes like the cultivation of soya, palm oil and other crops in an effort to sustain a growing population. These large-scale programmes have contributed vastly to raised carbon emissions as the trees that capture carbon dioxide and give off life-supporting oxygen are felled. The destruction of trees has also caused soil erosion on a massive scale, which has provoked landslides as dead roots can no longer provide a grip on the surrounding earth.

With the progression of time and improvements in technology, we have learned to harness the energy contained in other fossil fuels such as natural gas and oil. From the refining of oil, a carbon-emitting industry in its own right, we obtain petroleum, plastics and a whole range of chemicals and fertilisers, as well as the energy to drive power stations and other industrial processes. Natural gas is used by most of us to cook and heat our homes and without reliance by our industries on the many different uses of this material for manufacturing everyday products that we have come to rely on, our industrial base would collapse. It is now a fact of life that fossil fuels have become a very large part of our everyday living and are a cornerstone in our future survival plan. Unfortunately for us, the bad news is that they will not last for ever, as evidenced in Britain by the dwindling stocks of North Sea oil and the closure of coal mines (some of these on economic grounds and not always due to the depletion of coal). So what is the answer? In short, being more careful in the use of our precious resources and dramatically changing our throw-away culture to one of conservation and the recycling of all recoverable materials and waste, including food.

For a number of years, north London has been at the forefront of an innovative scheme to help turn what we might consider to be useless rubbish into a range of useful materials and also energy sources like gas and electricity. The project referred to was first commissioned in 1970 by the former Greater London Council (GLC) and built on land adjacent to the Eley's Industrial Estate, where the North Circular Road (A406) crosses the Lee Navigation. At the time of the project's conception it was known as the Edmonton Incinerator and largely dealt with the burning of household and other waste to produce electricity. After much scientific research and planning, this original scheme has evolved into LondonWaste's EcoPark. This considerably improved facility, with its team of passionate and dedicated staff, has totally transformed waste handling and recovery into something that would have been unimaginable only a few decades ago. However, as the author learned on a recent tour of the site, no member of staff is complacent and resting on their well-earned laurels as there is an ongoing review and monitoring of their systems, research and techniques to find new and improved ways of waste recovery. The ongoing aim is to reduce the reliance on landfill sites and also to help the world's carbon footprint become smaller.

The simplest way to describe the operation of LondonWaste's Lea Valley EcoPark is to break it down into its dedicated recycling and recovery areas.

Bulk Recycling Centre

This is where the recovery and separation of materials such as plastic, wood, fluorescent tubes, rubble, refrigerators and television receivers take place. After separation, the materials are stored and allowed to reach a required tonnage to ensure that the journeys to and from specialist contractors are kept to an absolute minimum. In turn, the contractors strip and process the various materials which will eventually be turned into new products.

Workman extracting wood for chipping from mixed waste.

Compost Centre

Green garden and kitchen scrap waste arrive at the centre and a mechanical shovel deposits the material into the hopper of a large shredding machine. Once shredded, the material is transferred by conveyor where water sprays dampen it before it is deposited onto a great heap within the centre. After standing for approximately 24 hours, the material is transferred into specially designed tunnels, each around 30 metres long. Here it stays for between two to three weeks as it undergoes the decomposition process, having controlled amounts of air forced through it. In the tunnels, each batch of material will heat up to the required temperature of 60°C which must be maintained for two consecutive days to meet government guidelines. The process, which effectively takes place in a container, is called "in-vessel composting". While the material is in the container, the temperature, moisture and air flow are carefully monitored. This is done from a control room that takes regular readings from temperature probes located inside the tunnels.

Moving green waste to the composting centre.

Compost Centre tunnels at London Waste.

London Waste's Compost Centre building.

Workman examining decomposing waste in containment tunnel.

After a period of between two and three weeks, the partially decayed material, which is now odourless, is removed and placed in another series of tunnels and the whole process is repeated. This ensures complete breakdown of the waste. After remaining in the tunnels for the required time, the material is removed to an outside bay where it is allowed to cool to around 40°C during the maturing process. This can last up to ten weeks. The last process is screening, where pieces of wood, plastic, stone or anything considered too large are removed by sieves. Remarkably, the Compost Centre is capable of handling 30,000 tonnes of green garden and kitchen waste a year.

The quality of the compost produced is such that it meets the British Standards Institution's specifications and is used by a range of different agencies, including farmers, landscapers and local authorities. As might be imagined, the Compost Centre and LondonWaste as a whole are required to meet strict inspection standards imposed by the Environment Agency, the Health and Safety Executive and the State Veterinary Service. So confident are the management of LondonWaste regarding the maintenance of these particular standards that it has, of its own volition, given the inspectorate security access passes to the site allowing them to arrive unannounced at any time of day or night.

Energy Centre

When all the recoverable waste has been separated for processing, the left over mixed material that would normally have gone to a landfill site is fed into massive furnaces where combustion takes place at a minimum temperature of 850°C, although temperatures are usually maintained somewhere between 900°C and 1,000°C. The heat generated by the furnaces is used to turn water into steam which drives turbines coupled to generators that produce electricity. During a typical year, enough electricity is generated to keep the whole site self-sufficient and the surplus, which is sufficient to power 24,000 homes, is fed to the National Grid. Flue gasses from the boiler chimney are scrubbed and passed through carbon and lime filters and are also subjected to electrostatic precipitators to remove harmful particles. Recently, an improved type of lime was introduced into the cleaning process, which has reduced considerably the need to mine large quantities of natural limestone. Materials that are known to produce high amounts of toxins are

LondonWaste Energy Centre, Enfield.

The control room is the hub of the Energy Centre, where all processes and emissions are monitored.

LondonWaste's Turbine Hall: these turbines are powered by steam, created by burning rubbish.

Mobile crane moving waste to boiler that will be burned to make electricity.

removed from the waste mix before they can be burned and the emissions from the chimney are strictly monitored by computers in the control room ensuring the air around the plant remains clean.

Clinical Treatment Centre

The Clinical Treatment Centre is run by a wholly owned subsidiary of LondonWaste Limited known as Polkacrest. It provides a safe and efficient clinical waste collection and treatment service for NHS trusts, private hospitals, nursing homes, dentists, doctors' surgeries and even tattoo artists. Perhaps some readers might remember the large factory-type chimneys that discharged liberal quantities of smoke and soot that were a by-product of the burning of clinical waste with coke (a combustible material, one of the by-products from the manufacture of town gas from coal). These chimneys graced the skyline behind many a Victorian hospital building and were accepted as part of the operation of the site.

Now that we know a little more about the harmful effects of pollutant discharges into the atmosphere, the current community should feel reassured that steps have been taken to have this unhygienic and haphazard method replaced with a modern efficient system of clinical waste treatment and disposal, contained within a specialist handling unit. Clinical waste is combusted at high temperature in a furnace and the resultant heat is used to produce steam from water to generate electricity. The process is similar to the way electricity is generated in the Energy Centre, mentioned above.

To ensure safe disposal and compliance with current legislation, LondonWaste operates a certification scheme for its customers. The certification scheme also applies to their service of sensitive document destruction.

Ash Recycling Centre

Ash left over after the mixed waste material has been subjected to high-temperature combustion in the furnaces to produce electricity is examined by a specialist team of operators who remove any residual metals before it is recycled and graded. The end product is used as a secondary aggregate by the construction industry and can end up in road-building schemes. This reduces the need to remove large quantities of natural materials from the earth like the gravels that once formed part of the Lea Valley after the ice sheet melted.

Wood Chipping Centre

Large quantities of wood, particularly old pallets, are shredded into chips which can then be used for weed suppression around shrubs, covering for garden paths, animal bedding, fuel and also new wood-based products. Perhaps the person fly tipping wood from a building site will end up purchasing quantities of particle board manufactured from the very waste of his illegal endeavours!

EcoPark Wharf

LondonWaste is ideally situated adjacent to the Lee Navigation and has its own wharf. It is the vision of the company to take waste management to the next stage and it is actively exploring environmentally friendly ways of using the canal system to move waste in bulk to and from the site. If this could be achieved, with the help of government agencies, it would reduce considerably the need for large-scale road transport movements, reducing congestion and further helping to improve our environment.

Future Possibilities – An Anaerobic Digestion Plant

The recycling of waste materials is always under scrutiny and responsible companies are constantly on the lookout for new technologies that will improve current efficiencies. One such technology, anaerobic digestion (AD), is a process under which biodegradable waste is processed in an enclosed reactor vessel, effectively a large tank that is devoid of air. As the waste breaks down under the effects of bacteria, biogas is produced. This is largely made up of a mixture of hydrogen, methane and carbon dioxide. The process also produces sludge, which after processing in the open air can be spread on land as a fertiliser. To maintain the AD process, the reaction vessel has to be heated. After initial priming has taken place, the biogas, created by the process, is then used to continue the heating of the waste and any surplus gas is used for the production of electricity. It will be recalled that this is the process that Guy & Wright have already adapted to power their Hertfordshire nursery.

While anaerobic digestion is a tried and tested technology which is becoming increasingly popular around the world with a number of industries, especially agriculture, horticulture and farming, it could prove costly to integrate such a process into the recycling of household waste. The biodegradable waste used is usually made up of vegetable and other food products. Therefore, it would be necessary to explore the economics and logistics of storing, collecting and delivering the material to LondonWaste from households within the seven North London local authorities (Barnet, Camden, Enfield, Hackney, Haringey, Islington and Waltham Forest) that currently make up the main waste catchment area. Otherwise, the cost to local council tax payers might prove unacceptable.

It is probably fair to conclude that rubbish recycling alone will not save our planet from the rapidly advancing effects of climate change. However, if nations around the world collectively began introducing some of the imaginative environmentally friendly initiatives that we have seen introduced in the Lea Valley region then we would really be getting somewhere. Those people who live in the Lea Valley region are fortunate to have the largest waste recycling plant of its type in Britain dealing with our daily creation of rubbish in some of the most environmentally friendly ways yet devised. This is helping to lower the earth's carbon footprint by reclaiming useful materials that would have otherwise gone to landfill. So perhaps the northern English saying "where there's muck, there's brass" should really apply in the south too!

REFERENCES

Author unknown, *Recycling North London's Waste*, LondonWaste Ltd, Edmonton

Author unknown, *EcoPark – Treating Rubbish as a Resource*, LondonWaste Ltd, Edmonton

Lewis, Jim, *Water and Waste: Four Hundred Years of Health Improvements in the Lea Valley*, Middlesex University Press, 2009

Scott, Nicky, *Composting: An Easy Household Guide*, Totnes, Devon: Green Books, 2005–6

Staff interviews, LondonWaste Ltd, January 2009

LEA VALLEY WILDLIFE HABITATS

We are extremely fortunate to have the Lee Valley Regional Park Authority (LVRPA) which has been doing its utmost to create and manage, with the help of recognised organisations and volunteers, the many wildlife habitats within the region for the last fifty years. These habitats, many rescued from the aftermath of sand, gravel and brick-clay extraction have encouraged the return of a diverse range of wildlife and plant species. This commitment by the LVRPA is having a major impact on the region's natural environment as it is helping many of the rarer species of flora and fauna to escape extinction's grasp.

Lee Valley Regional Park is a twenty-six-mile linear park that extends along the valley floor from Ware, Hertfordshire in the north to the River Thames in the south. On the way southward the park straddles the boundaries of Essex to the east and Hertfordshire and Middlesex to the west, providing a green lung for the citizens of London. There can be few capital cities around the world that can offer their citizens such a range of pleasant and healthy outdoor and indoor activities. These include the playing of various sports, walking, biking, boating and caravanning; not forgetting the opportunities for bird-watching and visiting the many museums and other visitor attractions that can be found around the region. Londoners are really spoiled for so much choice right on their doorstep.

Amwell Nature Reserve

Amwell Lane, Stanstead Abbots, Hertfordshire SG12 9SN

The reserve, managed by the Herts and Middlesex Wildlife Trust is the most northerly of the reserves within the LVRPA boundaries. It is also one of the last areas in the Lea Valley where gravel extraction took place. Over the past two decades, with a little help from the relevant agencies, nature has slowly claimed back much of its original homeland. The reserve is one of four sites within the park that together have been designated as the Lee Valley Special Protection Areas (SPAs) and which support numbers of internationally important wintering wildfowl. A patchwork of diverse habitats make up the reserve. These include lakes, grasslands, woodlands and water courses.

Hides provide the visitor with elevated views over the reed beds and, during the winter months, wildfowl such as Tufted Duck, Pochard, Gadwall and Shoveler can be seen. Other winter visitors include Smew and the elusive Bittern. In spring, waders appear and Little Ringed Plovers breed on site. They are regularly joined by other waders like Redshank, Green Sandpiper and occasionally rarer waders such as Dunlin.

Amwell Nature Reserve where gravel was once extracted.

Shoveler, one of the many water birds to be seen on the Lea Valley reserves.

The elusive Bittern.

During the summer months a Common Tern colony takes up residence on purpose-built rafts deployed on one of the lakes and, in a small meadow, marsh orchids bloom in early June. A boardwalk trail will allow the visitor to get up close and personal to nineteen species of the reserve's breeding damselflies and dragonflies. In the sky above, birds of prey including Buzzard, Sparrowhawk and Hobby may be seen.

Rye Meads

Rye Road, Hoddesdon, Hertfordshire ST12 8JS

Herts and Middlesex Wildlife Trust and the Royal Society for the Protection of Birds (RSPB) jointly manage this site which is made up of flood meadows, reed beds, lakes and wet woodland and forms part of the Lea Valley Special Protection Areas.

Rye Meads Nature Reserve.

In the summer months, a colony of Common Tern nests on rafts relatively close to the Gadwall hide and not far away an artificial bank has been created to attract nesting Kingfishers. The activities of these colourful birds can be watched from a dedicated Kingfisher hide. Nesting Kestrels, Long-tailed Tit and Reed Warbler occupy the reserve during the summer.

Wildfowl such as Shoveler, Gadwall and Tufted Duck inhabit the lakes in winter and in the wet scrape areas

*A pair of Common Tern on
Rye Meads Nature Reserve.*

*A Kingfisher with a catch at
Rye Meads Nature Reserve.*

Green Sandpipers dabble for food. It is possible for a fortunate visitor to spot the Bittern in the reed beds. These birds are now becoming familiar visitors to the Lea Valley reserves.

The Otter, another illusive animal, which had become extinct in the region and was reintroduced to the Amwell Nature Reserve some years ago, has left spraints (faeces) which show that this animal's nocturnal hunting territory now includes the reserve.

To help conserve rare plant species such as the Meadow Rue, Konik Ponies have been introduced to graze the wet meadow areas. Water Buffalo were also used in the past but when the tough grazing was done they were retired. Controlling the vegetation in this way has encouraged wading birds like the Lapwing to nest.

*Stanstead Innings Nature
Reserve.*

Stanstead Innings

*Marsh Lane, Stanstead Abbots,
Hertfordshire SG12 8HL*

This reserve is solely managed by the Lee Valley Regional Park Authority which has carried out extensive reed-bed improvements and also maintains a continuing programme of habitat improvement. The boardwalk provides a pleasant viewing platform and compliments the Shoveler Hide where Gadwall, Tufted Duck and Shoveler may be observed. Looking from the Sandpiper hide to the shallow water areas Teal, Lapwing and Green Sandpiper can be seen feeding in the margins. Also, visitors to the hide in winter months may be rewarded by the sight of

Bittern roosting in the reed beds, as these birds regularly use the site during this time of year.

Late spring brings the wildflower meadows to life and it is a joy to see blooms of the Ragged Robin and Bee Orchids lighting up the world in the early sunshine. The flowers offer a colourful welcome to the different species of warblers which include Reed and Sedge.

Ragged Robin among other wild flowers on Stanstead Innings Nature Reserve.

Glen Faba

Glen Faba Road, Nazeing, Hertfordshire CM19 5EX

This reserve consists of two lakes fringed by reed beds and is surrounded with wooded clumps and wildflower meadows. Wigeon, Gadwall and Tufted Duck can be seen on the open water areas. Glen Faba's largest wooded island supports a small colony of Cormorants and is also home to a heronry. Throughout the year, on the lake called the Stort Pit, Little Grebe may be seen. In the summer months the surrounding reed beds provide habitats for Reed and Sedge Warbler and also Reed Bunting, which frequent the site throughout the year. Take a stroll on a warm summer evening and you could witness bats performing their aerobatic manoeuvres as they search for their supper.

A Wigeon on Glen Faba Nature Reserve.

Glen Faba Nature Reserve with power station in the background.

A Little Egret looking for lunch.

Glen Faba Nature Reserve.

Admirals Walk Lake

Admirals Walk, Hoddesdon, Hertfordshire EN11 8AB

The River Lynch flows along the northern boundary of this twenty-five-acre shallow spring-fed lake bordered with wooded areas. This combined area of water and woods provides the perfect habitat for a variety of insects which include Dragonfly and Black-tailed Skimmers. All may be seen during the summer months. The White-legged Damselfly once frequented this site but has not been seen for some time. This could be a warning of how delicately nature is balanced!

From April to October, particularly on warm evenings, the reserve becomes the feeding ground for Pipistrelle and other species of bats. The rarer species, Daubenton's Bat, named after the French naturalist Louis-Jean-Marie Daubenton (1716–1800), can be seen hunting for insects as it skims close to the water. This characteristic hunting habit distinguishes it from the Pipistrelles which tend to fly higher and do not always frequent wetland habitats. Other distinguishing features of Daubenton's Bat, over the Pipistrelle, is that the former has a reddish-pink face and nose, and no fur around the eye area.

Coot and Little Grebe can be seen on the open water areas throughout the year and during the summer months the reserve becomes a popular foraging area for Swallow and Sand Martin.

Nazeing Meads Nature Reserve.

Nazeing Meads

Dobbs Weir Road, Nazeing, Essex EN9 2PD

Here advantage has been taken of the River Lea flood relief system and the three large settlement lagoons now provide a welcome food source for wildlife. Diving birds including the Tufted Duck revel in foraging beneath the deep water. From the bridge across the flood relief channel, wintering duck including Goldeneye and Goosander may be seen. Also during winter months there is normally a large roost of Common and Black-headed Gulls and on occasion the rarer Mediterranean Gull might be sighted.

Silvermeade

Mill Lane, Broxbourne, Hertfordshire EN10 6LX

This site provides the ideal habitat for the endangered water vole, an animal immortalised as Ratty in the children's book, *Wind in the Willows*. In the 1920s, the North American mink was introduced to Britain and bred on special farms for its fur. Fur was a popular material, particularly in women's fashion, and by the mid-1950s several hundred fur farms were thriving in the UK. As is inevitable, some of the animals escaped and others were released by well-meaning animal rights activists who clearly had no idea what devastation their actions would wreak on the indigenous wildlife. Mink are avaricious hunters and will take smallish animals, the water vole being a favourite dish of the day. They will also take birds, frogs, fish and bird eggs. However, the good news is that surveys carried out by ecologists from the Canal and River Trust has shown that our carefully managed nature reserves are helping the water voles make a comeback. At the same time, wild mink numbers must be monitored and the animal kept under control. Some naturalists have suggested the increasing otter population might be playing a part in this.

Silvermeade Nature Reserve.

The Silvermeade site consists of wet meadows that are criss-crossed by a series of ditches, pools and reed beds which make the area ideal for dragonflies and also the endangered water vole. With patience, this attractive creature may be viewed in spring as water voles tend to re-examine their territories around this time of year. Clues to their existence can be seen throughout the year as it is possible the visitor might spot droppings and also pathways running through the vole's neatly cropped grass and along the banks of ditches.

Water Vole on Silvermeade Nature Reserve.

Spring is a good time to see the delicate pink–white Cuckooflower which blooms around the time when the bird is first heard, hence the name. Also about this time, amongst the grasses and sedge in the wet meadow areas, the Ragged Robin shows its beautiful pink flowers.

The earliest of the dragonflies to emerge is the hairy variety which can be seen from May onwards hunting for food above the site's numerous ditches. Later in the year these dragonflies are joined by the Banded Demoiselle sporting a glittering blue body; the male of the species is identified by the dark banding on the wings.

Grass snakes, the largest of our British species and non-venomous, are often seen hunting in the long grass along the edge of water channels. Being good swimmers their main meals are normally amphibians such as frogs, toads and newts, although they have occasionally been known to take small mammals and birds. Their natural enemies are badgers, hedgehogs, foxes and also birds of prey, although the domestic cat can be a problem to this protected species.

Cuckoo Flower on Silvermeade Nature Reserve.

Broxbourne Old Mill & Meadows

Mill Lane, Broxbourne, Hertfordshire EN10 6LX

Broxbourne Mill is first mentioned in the Domesday Book – a mill has stood on this site for over nine-hundred years. Milling of grain ceased in 1891 and afterwards the mill had various uses until it was destroyed by fire in 1949. In the late 1970s the Lee Valley Regional Park Authority restored the waterwheel by replacing the paddles with recycled plastic replicas. Apart from helping to reduce the cost of maintenance, the LVRPA restoration work will help to remind us and future generations of Broxbourne's industrial past.

Clumps of Giant Horsetail, a peculiar looking plant that can reach around 1.5 metres (5 feet) tall, sometimes referred to as the 'living fossil'. The Kingfisher and Grey Wagtail are regular visitors to the millstream and pool area.

Rusheymead

Mill Lane, Broxbourne, Hertfordshire EN10 6LX

This site is a mixture of open grassland and scrub with areas of mature woodland. During the summer month's flocks of small birds are attracted to the site, probably as the scrub and woodland provide safe cover from hunting Sparrowhawks. The summer is a good time to see Warblers and Bullfinches which can be viewed throughout the year. On the grassy areas, Green Woodpeckers, a ground-feeding bird, can be seen searching for ants and their eggs which are a favourite meal.

Fishers Green

Stubbins Hall Lane, Crooked Mile, Waltham Abbey, Essex EN9 2EF

This site forms part of the Lee Valley Park Special Protection Area (SPA) and is the centrepiece of Lee Valley Country Park where the Turnford and Cheshunt Pits provide an important haven for wintering wildfowl. The site's Seventy Acre Lake has seen extensive improvements to its reed beds as the LVRPA wish to attract visiting Bitterns to take up permanent residence and allow them to breed.

A relative of the heron family, the Bittern, according to the Royal Society for the Protection of Birds (RSPB), is making a comeback in the UK after almost reaching extinction. In the Middle Ages, the bird was a delicacy at banquets and later in the eighteenth and nineteenth centuries the birds' natural habitats were destroyed as polluting discharges occurred as the region became more industrialised. Also the extraction of sand and gravel for building and civil engineering projects destroyed reed beds and water courses. If the LVRPA habitat-regeneration programme works, then in future springs the booming mating call of the male Bittern will be herd like a ship's foghorn across the Lea Valley. Unfortunately, during the winter months, the Bittern is not easy to see as the bird's stripy brown plumage blends perfectly with the reeds when the bird is hunting at the water's edge for fish and amphibians.

Through the year, the reed beds provide habitat for other birds like Snipe, Lapwing and also the illusive Water Rail. Islands on the site's lakes also attract Snipe

Fishers Green Nature Reserve.

The illusive Bittern, Fishers Green Nature Reserve.

A Tern posing for the camera.

Bittern hunting in the reeds at Fishers Green Nature Reserve.

A Smew on a visit to Fishers Green Nature Reserve.

Roesel's Bush Cricket.

Orchids, not early marsh orchid.

and Lapwing and wintering wildfowl like Gadwall, Smew, Shoveler, Goldeneye and Goosander; and on rare occasions, the Pintail can be seen on the lakes. During the summer months, a Common Tern and Black-headed Gull colony takes over the rafts on Seventy Acre Lake.

In spring and early summer, particularly at dawn or dusk, the visitor might hear the beautiful song of the Nightingale when strolling past the wooded area north of Fishers Green car park. Keeping a watchful eye open along Hooks Marsh Ditch, close to the Bittern Information Point, Water Voles may be seen going about their business. Also be on the lookout for an iridescent flash of blue above the watercourses as Kingfishers are commonly seen in this area.

Goosefield

Stubbins Hall Lane, Crooked Mile, Waltham Abbey, Essex EN9 2EF

Goosefield Nature Reserve, as seen from the hide.

To attract different species of bird and other wildlife to an area, there is a need for a range of habitats to suit different individual hunting, nesting and feeding habits. Fortunately the Lee Valley Regional Park Authority, with its partners, has an ongoing programme to create such places. Goosefield is a particularly good example as the wet meadows punctuated with shallow pools and ditches provide excellent habitat for waders and grazing wildfowl. However, despite all the good intentions of the partners to develop new wildlife habitats, progress has been delayed by an unwelcome outsider.

New Zealand Pigmyweed (*crassula helmsii*), a non-native invasive plant from Australia and New Zealand, has been discovered on this site. The plant grows vigorously in our waterways choking the system and out-competing native plants. This has caused the LVRPA to curtail its habitat management programme by reducing the amount of water onto the site until the problem has been satisfactorily controlled. There are suggestions that the invader was first discovered in Britain in the 1970s residing in a roadside pond close to houses in the New Forest area of Hampshire and has since spread to other sites across the UK. However, further research suggests that the plant first came to Britain as early as 1911 and there are other reports of the plant being grown at an Enfield nursery in the 1920s.

The plant can be easily spread by tiny fragments being carried in flowing water and contamination can also be distributed by vehicles, machinery, birds, animals and people.

While this is not a happy situation to be contemplating in the year 2017, it is certainly a wake-up call to all of us who wish to see nature returned to its former glory after man's un-thoughtful interventions. We must also be exceptionally vigilant with regard to the plants that we select for our gardens. Government agencies across the world must also ensure that effective and watertight controls are in place to prevent nurseries, garden centres, aquatic centres and unthinking people from importing any type of plant that could be harmful to their country's indigenous flora and fauna. Currently, it is an offence to allow crassula to grow in the wild. However, my Lee Valley Regional Park contacts have pointed out that "new invasive species are arriving all the time". This might appear as an alarming revelation but at least it should give us a large amount of confidence that a responsible body is always watchful and is monitoring the situation on our behalf – although this should not prevent us as individuals from being exceptionally vigilant.

Holyfield Lake

Stubbins Hall Lane, Crooked Mile, Waltham Abbey, Essex EN9 2EF

Holyfield Lake Nature Reserve, picture taken from the hide.

Throughout the year, the lake is home to Great Crested Grebe, Little Grebe and Coot. From the path that goes towards the Grebe Hide, the visitor will see a Cormorant roost and also a heronry. When walking through the River Lee Country Park, the visitor is sure to come across many clumps of alder as the tree is prevalent in damp and boggy areas. In winter, large flocks of Siskin and tit can often be seen in the upper canopy feeding on the plentiful supply of alder seeds.

Hall Marsh Scrape

Fishers Green Lane, Crooked Mile, Waltham Abbey, Essex EN9 2ED

Hall Marsh Scrape.

The site consists of four shallow artificial scrapes and gravel islands that have been designed to attract waders. In a world where many animals and plants are on the verge of extinction, it is nice to see that organisations like the LVRPA and its partners are helping to bring about programmes for species diversification across the region. Here on Hall Marsh Scrape, we see another example of the design and recreation of the lost habitats that have been destroyed through man's lack of understanding that a fragile balance exists between humans and nature. Although there are still deniers with regard to how our activities affect the climate, fortunately there is a growing scientific consensus that would suggest that reducing harmful emissions, restoring lost forests and hedgerows, and generally respecting our environment form the most sensible long-term approach to protecting our planet.

In the spring, the new shingle islands provide the ideal habit for the Little Ringed Plover, returning from Africa to breed. Other regular visitors to the Scrape are Common and Green Sandpipers as they pass through the valley on their long migration from Northern Europe to Africa. These waders were once residents of the region and it is hoped that by providing them with the right kind of habitat they will be encouraged, once again, to take up permanent residence.

During the summer months, Common Tern can be seen overhead as they fly from their colony on Seventy Acres Lake and Hobbys can also be seen hunting for their favourite food as dragonflies dart energetically above the adjacent flood relief channel.

In winter, the site is popular with Wigeon, Gadwall and Shoveler and large flocks of Lapwing and occasionally Golden Plover can be seen. Rare sightings of Pintail, Spoonbill and Spotted Crake have been made which suggest that the Scrape should be a location worth a visit.

Lee Valley Park Farms

Stubbins Hall Lane, Crooked Mile, Waltham Abbey, Essex EN9 2EF

There are two farms on this site – Hayes Hill Farm and Holyfield Hall Farm – each giving the visitor a different experience. Hayes Hill Farm has been designed to introduce children to a wide range of domesticated animals which include cattle, sheep, pigs, goats, rabbits, guinea pigs and chickens has proved to be a popular family day out for many years.

Great days out at Lee Valley Park Farms – something for the whole family.

Holyfield Hall Farm is a mixed dairy, beef and arable farm that demonstrates quite perfectly how farming and wildlife can co-exist. Field margins have been encouraged as their variety of plants and grasses provide habitats for insects which provide food for nesting Whitethroats and Yellowhammers in the nearby hedgerows. Small mammals also flourish in the margins and attract Kestrels, Barn Owls, Little Owls and Sparrowhawks. During autumn and spring, migration Wheatears can regularly be seen around the field margins, while Fallow and Muntjac Deer are present on the farms.

West Side Gravel Pits

Windmill Lane, Cheshunt, Hertfordshire EN8 9AJ

These water-filled gravel pits, and the grassland areas around them, located west of the Lee Navigation, provide a place where rare orchids thrive. When ash from the old coal-burning Metropolitan Power Stations, particularly those at Brimsdown, had to be got rid of, the nearby water-filled pits provided the ideal place for the waste to be dumped. At the time, nobody could have predicted, or would have given a thought, to the effect this might have on the surrounding grassland areas. It would seem that water from within the pits has allowed the ash to leach nutrients into the surrounding soil which has accidentally encouraged the growth of a variety of orchids, particularly in the North Metropolitan and Bowyers Water areas. These include: Early Marsh, Southern Marsh, Common Spotted, Pyramidal, Twayblade and Bee Orchids. In the summer, Bowyers Water also has a magnificent display of Water Lilies.

A great summer-month attraction can be seen, or perhaps more correctly heard, in the areas of Thistly and Cheshunt Marsh, at a time when grasshoppers and crickets become their most active. Look out, or better still listen, for the characteristic hissing sound (which some have likened to electricity pylon insulators fizzing during damp weather) of Roesel's Bush Cricket. The insect is named after the German miniature portrait painter and naturalist, August Johann Roesel von Rosenhof (1705–59). Also during this time of year watch out for the different varieties of butterflies, particularly the Speckled Wood, which flitter regularly across woodland paths on sunny days.

Cornmill Meadows and Tree Park

Crooked Mile, Waltham Abbey, Essex EN9 2ES

Cornmill Meadows reserve, within the River Lea flood plain.

There are few remaining examples of natural flood plain within the boundaries of the Lee Valley Regional Park, but Cornmill Meadows can claim semi-natural status as it has avoided the devastating effect of the gravel extractors' machinery. The area is crisscrossed by a number of waterways, ditches and pools which attract a variety of wildlife throughout the year. Hay meadows and woodland complement the site and in the tree park there is a network of glades and paths to explore. The site is located just east of the Royal Gunpowder Mills (see the chapter 'Tourism – the Lea Valley's Alternative Economy'), an attraction well worth a visit as part of the Mills complex is a nature reserve in its own right. Cornmill Stream, part of the River Lea network, runs through the meadows and once powered the mill and filled the fish ponds. The remains of both can still be seen within the precincts of the Abbey.

A Lapwing visits.

Over half of the dragonfly species found in the UK can be seen at this site, notably the White-legged Damselfly and Hairy Dragonfly. A stroll through

A Hairy Dragonfly at Cornmill Meadows.

the wooded glades on a mild evening in late summer to view large numbers of Migrant Hawker Dragonfly can be particularly rewarding.

The spring and autumn seasons are the best times to see species like the occasional Ruff and Black-tailed Godwit, alongside the more frequent Redshank and Common and Green Sandpiper. In winter, large flocks of Teal and Wigeon can be viewed which are often joined by even larger flocks of Lapwings and occasionally Golden Plover. As might be expected at a dragonfly reserve, Hobbys are often spotted hunting their prey.

A visit from a Green Sandpiper.

Sewardstone Marsh

Sewardstone Road, Near Chingford, London E4 7RA

Sewardstone Marsh is the site of the former Knight's gravel pits which have been in-filled with water. The Marsh also includes the wet grassland area of Patty Pool Mead where Snipe can be seen feeding in the winter and Long-eared Owl have recently been spotted using the Marsh at this time of year. Scrub clearance along the site's ditches has improved the habitat for Water Vole which may be why the Long-eared Owl has recently been seen.

In the summer, the beautiful song of the Nightingale may be heard and, if fortunate, the visitor might catch a fleeting glimpse of this elusive

A Snipe on Sewardstone Marsh Nature Reserve.

bird. The Nightingale is a smallish bird about the size of a House Sparrow. It has a light-brown back and a buff-to-white breast and underbelly. The bird tends to hide in dense scrub and woodland areas.

At the northern end of the Marsh, west of the flood relief channel, is Enfield Island Village. This is a relatively new mixed housing development with a small retail and commercial centre. The centre has been sensitively constructed around a restored and modified Grade II listed building which was once a large machine room. Rents for the various businesses now established within this building are collected by a not-for-profit company, RSA IV. The surpluses from these rents are transferred to the RSA Trust, a registered charity that manages the site and also funds local community start-ups and distributes grants in support of good causes.

The Enfield Island Village stands on the site of the former Royal Small Arms Factory (RSAF), the birthplace of the famous Lee Enfield Rifle. Prior to the opening of Enfield Island Village in 2001, it had been planned to open up the old barrel grinding mill headstream and barge turning circle that had remained, for many years, in-filled with rubble and concreted over. When the work was completed, a narrow boat was placed on the new water feature to make the connection between the small arms factory and the nearby Lee Navigation that had been used for the transportation of goods and materials. Reeds were planted in the old barge turning circle and the pond soon became a haven for wildlife, attracting Swans, Mallards, Moorhen and Coots. Common birds like Pied Wagtails, Blackbird, Finches and Tits can be seen feeding around the pond's margins. In the summertime, dragonflies and damselflies dart and skim across their new artificially created pond.

The RSA Trust operates a small Interpretation Centre which is located under the nineteenth-century Italianate clock tower which was part of the old RSAF large machine room. Visitors wishing to view the Centre's displays should enquire at the RSA Trust's on-site office about a fact-finding accompanied visit. Here we have a prime example of how a modern building development, on a former brown-field site, can sensitively be planned to take account of the history of the place while creating a scheme that is visually appealing to both residents and visitors and is also an attraction to wildlife.

Rammey Marsh

Ordnance Road, Enfield, Middlesex EN3 6TH

Entrance to this rough grassland site, which is intersected by a ditch from the Small River Lea, can be accessed from the Lee Navigation towpath. The site

Pyramidal Orchid on Rammey Marsh Nature Reserve.

Bee Orchid on Rammey Marsh Nature Reserve.

ditch is connected to a seasonal pool and both of these features support Water Voles and Grass Snakes. At the northern end of the site, in the months of May and June, a large colony of Bee Orchids form a glorious display that will not fail to please the visitor.

Rammey Marsh Nature Reserve.

Ponders End Lake

Lee Valley Golf Course, Meridian Way, Edmonton, London N6 0AR

As the address might suggest, Ponders End Lake, which has a surface area of one acre, is a wildlife habitat in a rather unique setting. A path leads to a hide that can be accessed from a gate onto the golf course near the Lee Valley Leisure Complex. From this vantage point, look towards the reed-fringed areas of the lake to see Reed and Sedge Warblers during the summer months and Reed Buntings all year round. Common Terns, returning from their long flight from Africa, nest on the gravel surface of the lake's island.

Reed Warbler at Ponders End Lake Nature Reserve.

Lapwings are often seen using the island and, in the winter, Wigeon may be viewed on the lake and can be seen contentedly grazing on the golf-course fairways. Kingfishers are occasionally known to flash by and can usually be distinguished from the flying golf ball by their bright ultramarine colour!

Chingford Reservoirs

Lea Valley Road (A110), Chingford, London E4 7PX

The Chingford Reservoirs are made up of the more northerly King George V and the William Girling. They are divided by the Lea Valley Road which is one of the few crossing points across the valley. Often when driving or walking along this road, sheep can be seen grazing on the reservoir embankments, acting as environmentally friendly lawnmowers. Interestingly, the William Girling is built on the site of the former Chingford Airfield that was used to train Royal Navy Air Service (RNAS) flyers during the First World War. The reservoirs are owned by Thames Water and have been designated a Site of Special Scientific Interest (SSSI) due to their importance for over-wintering wildfowl. During the harsher months of the year, roosting Gulls can number up to 50,000. These include species like Common, Herring, Black-backed and Black-headed.

A rare appearance by an unlikely visitor to the King George V Reservoir during the First World War. These were known as 'kites'!

Waterfowl which consist of Teal, Goosander and Goldeneye over-winter on these reservoirs and the site is nationally recognised for wintering Black-necked Grebe. Rarer species of wildfowl can be brought into the area, particularly at times when harsh weather hits the European continent.

It is worth remembering that access to these reservoirs, because of deep-water dangers, is not always possible so it is worth contacting Thames Water to find out the current position.

Tottenham Marshes

Watermead Way, Tottenham, London N17 0XD

The Lee Navigation crossing Tottenham Marshes.

Peregrine Falcon on Tottenham Marshes.

Tottenham Marshes have changed out of all recognition since the 1950s. Once, piles of dumped cinder and clinker (the residue from industrial boiler furnaces) and other detritus littered the landscape, which also accommodated patches of bare soil, the tell-tale sign of a bicycle dirt track. On these compacted areas of earth, young boys, including the author, sped around madly on unsafe machines in pursuit of the new cycle speedway craze, trying to emulate their motorcycle heroes. Fortunately, through sensitive conservation and planting programmes, the Marshes are now a large expanse of grassland with wildflower meadows where Bee Orchid numbers are annually increasing alongside those of the rarer Wall Bedstraw.

It is worth looking up towards the top of the electricity pylons that cross the Marsh as Kestrel's use these striding giants to perch on while watching for

Stonechat on Tottenham Marshes.

movement of small mammals below. Sand Martins have been encouraged to visit and nest on the Marsh during the summer months. Holes drilled into the concrete-wall culverting of Pymmes Brook provide the ideal home for these birds, the smallest of the Swallow family, that have taken the long journey from Africa to reach the UK to breed. In winter, the Marsh is also a good place to see many small birds, particularly Linnets which arrive to feed on seeds of Teasel, Dock and Thistle.

Insects are doing well on the Marsh, particularly the Wasp Spider, so called for its distinctive wasp-like markings. In recent years, their numbers have increased in Britain due mainly to improved summer temperatures. The species, originally an inhabitant of the Mediterranean, was first identified in the UK in the early 1920s in the warmer area of Britain's south coast. The marsh grassland provides the ideal habitat for this spider to lay its eggs and make its unique web that has a characteristic zigzag pattern down the middle in which grasshoppers and small insects are trapped. Female spiders are around three-times larger than the male and possess a nasty bite which can make mating a little difficult for their partners, to say the least. During the mating process it is not unusual for the male to end up bitten and wrapped in silk to provide a meal for the female – not a good start for any kind of relationship! The milder weather has encouraged the Small Red-eyed Damselfly into the area. They can be seen in late June to September delicately balanced on floating vegetation on the nearby Lee Navigation.

Walthamstow Marshes

Lea Bridge Road, Leyton E10 7QL

The Walthamstow Marshes are one of the few areas in the Lea Valley which can be claimed as Lammas Land and where the annual tradition of beating the bounds still takes place. Local people were once allowed to cut hay and graze their animals on these marshes. In 1981, the Marshes were granted the status of a Site of Special Scientific Interest (SSSI). To maintain the unique grassland areas, the LVRPA introduced cattle to manage the Marshes in the way it was traditionally done by local people. This will hopefully conserve the flora and fauna of the area for future generations to enjoy. It would appear that this plan is working as it has been recorded that the Marshes support 350 different plant species. This includes the rare Creeping Marshwort with

Walthamstow South Marsh in winter.

Creeping Marshwort on Walthamstow Marshes.

A rare Essex Skipper on Walthamstow Marsh.

Belted Galloway grazing on Walthamstow Marsh.

its tiny flowers which can be seen along the grazed edges of draining ditches. Only on Walthamstow Marshes and on two other sites in the UK is this rare plant known to grow.

Seventeen species of butterfly breed on the Marshes including the rare Essex Skipper. In the summer Willow, Sedge and Reed Warblers arrive to breed and join the indigenous Reed Bunting. Water Voles colonise the drainage ditches and several species of small mammals populate the grassland areas, making the place an attractive hunting ground for Kestrels.

In July 1909, the Marsh provided the take-off site for Alliott Verdon Roe (of AVRO fame) who became the first Briton to fly in an all-British aircraft. His triplane (three wings) was constructed in two arches of the viaduct that carries the Liverpool Street to Chingford railway line across the River Lea. A plaque on the viaduct wall commemorates the occasion.

Waterworks Nature Reserve

Lammas Road, off Lea Bridge Road, Leyton, London E10 7NU

Waterworks Nature Reserve.

This reserve was built on the site of the former Essex Filter Beds that were part of the old Lea Bridge Waterworks complex. It boasts one of the largest bird hides in London where visitors can look out for a variety of wildfowl from the central hide including Gadwalls and Shovelers. In the summer months, Little Grebes and Pochards breed on site and winter is a good time to see Teal and Snipe. In spring and autumn, passing waders like Common and Green Sandpipers regularly visit the site. Occasionally, the site has been known to attract rarer visitors such as Black-tailed Godwits and Wood Sandpipers.

Tufted Duck at the Waterworks Nature Reserve.

In recent years, artificial nesting towers have been constructed for visiting Sand Martins which the birds were originally grateful to call home. Each year the number of nesting pairs increased. However, in recent years the birds have not returned but the towers have been left in place just in case they change their minds. The towers are made out of large sections of concrete pipe drilled with holes around the perimeter and then stood vertically on end. An artificial nesting bank for Kingfishers

has also been constructed on site to see if these birds will take up the offer of an artificial home!

If you are an enthusiastic bird watcher, the Waterworks Nature Reserve is the place to go. From the seclusion of the hides, birds like Garden Warblers, Willow Warblers, Chiffchaffs, Blackcaps, Whitethroat and Sedge and Reed Warblers are amongst those that can be seen.

Pochard at the Waterworks Nature Reserve.

Middlesex Filter Beds

Lea Bridge Road, Leyton, London E5 9RB

This reserve, a short distance from the Waterworks Nature Reserve, was constructed on the site of the former East London Waterworks Company (ELWC). Coincidentally, the ELWC built and owned the reservoirs off Ferry Lane which have now become the Walthamstow Wetlands nature reserve and the inspiration for this book.

Of all the nature reserves within the Lee Valley Regional Park, the Middlesex Filter Beds are one of a few that have left clues to their industrial past. Several of these clues have now been transformed into works of art. Those of us with an interest in industrial archaeology can marvel at the great granite blocks with their holes and grooves, now arranged as artwork around the reserve. The blocks were rescued from the site's old engine house when it was demolished and were probably the mountings for the massive iron flywheels and beam engines that were once used for pumping water. When the waterworks was built in the mid-nineteenth century, the Victorian engineers and scientists were desperately searching to find sources of cleaner water located as far as possible from the contaminated supplies that had previously been taken from the River Thames.

View across the Middlesex Filter Beds Nature Reserve.

The reserve has a variety of habitats which makes it a popular place for wildlife viewing throughout the year. Wooded areas attract flocks of finches and tits while the remains of industrial brickwork provide shelter for amphibians. In the spring, the wetland areas are home to toads, frogs and newts. Green and Great Spotted Woodpeckers are often seen on site as well as Sparrowhawks and Kestrels that come in after hunting on the nearby Hackney Marsh. The waterway known as Middlesex Filter Beds Weir, which splits the Lee Navigation from the loop of the

River Lea that flows across Hackney Marsh, is a good place to see Grey Wagtails and Kingfishers that nest along the river banks. Take a close look at the weir and an Eel pass will be seen that allows Eels to swim further up the River Lea. When the Eels have matured, the Eel pass allows them back to continue their long journey across the Atlantic to their traditional breeding ground of the Sargasso Sea, south of Bahamas.

Over two-hundred species of plant have been recorded growing on site including Purple Loosestrife and Cuckooflower and several different mosses and liverworts have taken up residence along areas of old brickwork.

Middlesex Filter Beds Nature Reserve, constructed on the site of the former waterworks at Lea Bridge, Leyton.

Bow Creek Ecology Park

Wharf Side Road, off Bidder Street, Canning Town, London E16 4ST

Wild flowers grow at Bow Creek Ecology Park where industry once prospered.

Bow Creek Ecology Park, nestled in the meandering loop of Bow Creek, was once an osier bed where willow 'withies' were grown for making baskets, fish-traps, household furniture and also baby carriages (prams). Late-nineteenth-century maps show the site sitting adjacent to an industrial complex with several wharfs, a sack and bag works, an oil refinery, a galvanised ironworks and oil mills. This would have been an extremely unhealthy area to work and far removed from today's little East End wildlife haven that visitors can enjoy.

Bow Creek Ecology Park, a jewel in east London.

In summer, take the path through the wildflower meadow to see different species of butterfly that have been attracted to the nectar-producing plants. Orange Tips, Green-veined Whites and Small Coppers are regularly seen. Between June and August, the magnificent Emperor Dragonfly's emerge and at first both male and female are similarly coloured green, but later the male sports a sky-blue abdomen.

It is worth keeping a lookout over the tidal mudflats that surround the park as flocks of Redshanks often take the opportunity to feed when the River Thames is on the ebb. Kestrel can be seen hunting along the verges of the Docklands Light Railway that crosses the site. Reports of Black Redstarts in the area have been received by LVRPA. These birds tend to favour habitats like derelict industrial sites.

East India Dock Basin

Orchard Place, Canning Town, London E14 9QS

As the name suggests, this was once the property of the East India Company that built the docks in the early nineteenth century. From here, East Indiamen sailed with goods for the East India Company and returned with cargoes of silk, tea and spices. During the Second World War, the dock was used to construct the floating Mulberry Harbours that were successfully used in the 1944 D-Day Landings in France.

The Dock Basin was once served by giant lock gates which, when closed, protected the penned-in water from being lowered by the falling tide on the River Thames. This allowed ships to be loaded or unloaded while being kept afloat.

Now the Dock Basin, with its brackish water, acts as a tidal lagoon and is one of the more unusual London nature reserves. At the northern end there are mudflats with a small salt-marsh area supporting Common Reed, Sea Club-rush, Buttonweed, Sea Milkwort, Sea Arrowgrass and Wild Celery. The grassland area supports many native flowers including Ladies Bedstraw and also some other interesting plants like Salsify and Warty Cabbage.

Urban habitats, East India Dock Basin.

East India Dock Basin Gates.

Urban habitats, East India Dock Basin.

Common Teal at the East India Dock Basin Nature Reserve.

Every summer a colony of Common Terns returns to the artificial rafts on the lagoon. Cormorants can also be seen throughout the year in characteristic pose with wings open and half-folded, catching the early sun. In the winter months large numbers of Shelducks and flocks of over 150 Teal can be seen. Black Redstarts often visit the site especially in spring and autumn. In the bordering scrub, a Barred Warbler was recently recorded.

It is worth taking binoculars as the site provides a good vantage point to look across the River Thames towards the O2 Arena. Peregrine Falcons and a wide variety of gulls can be seen on a good day in the area of the O2 Arena's support towers.

REFERENCES

Author unknown, *Where to Watch Wildlife in the Lee Valley*, Lee Valley Regional Park Authority, 2012

Bromberg, Stephen, Head of Communications, LVRPA, personal conversation, December 2016

Lewis, Jim, *London's Lea Valley: Britain's Best Kept Secret*, Phillimore & Co. Ltd, 1999

Lewis, Jim, *London's Lea Valley: More Secrets Revealed*, Phillimore & Co. Ltd, 2001

Patrick, Cath, Senior Conservation Officer, LVRPA, personal conversation, December 2016

Note

All the images used in this chapter were provided by:

Lee Valley Regional Park Authority
Hertfordshire & Middlesex Wildlife Trust
Brenda Chanter
Brian Anderson
Dennis Meadhurst
Graham Canny
Ken Bentley
Mark Braun
Paul Lister
Steven Swaby
Tim Hill

THE LEA VALLEY'S SECRET FISHERY

I wonder how many people who pick up this book will have heard of the Amwell Magna Fishery. Until late 2016, the author had no knowledge of the place despite his thirty-plus years of researching the Lea Valley region for interesting stories for his twelve published books; Amwell Magna Fishery never appeared on the radar. Perhaps this was due to a focus on industrial firsts, but nevertheless the story of the fishery, with its connections to the wider ecology of the Lea Valley region, is too good to miss.

In his book *The Compleat Angler*, first published in 1653, Izaak Walton (1593–1683) mentions fishing the River Lea near Amwell Hill. As the Act of Parliament to construct the Lee Navigation had not been passed until 1767 (after the 1766 report of the civil engineer John Smeaton), Walton could only have known a River Lea that followed its natural course. However, Walton would have been aware of Sir Hugh Myddelton's New River, dug between 1609 and 1613, that took fresh drinking water to London from the springs at Amwell and Chadwell. So it is possible that Walton fished the loop of the old River Lea which now forms the Amwell Magna Fishery – although without documentary evidence we cannot be sure. Even if he did not fish this actual spot, it would seem inconceivable that he did not know of it.

The Amwell Magna Fishery is the oldest fly-fishing club in Britain and still owns and fishes the same two-mile stretch of the River Lea, often referred to locally as the loop, that flows between the town of Ware and the village of Stanstead Abbotts. The fishery was founded in 1841 by William Shackell, who, according to Feargal Sharkey, the present chairman and historian of the fishery, "led a colourful if not turbulent life. As a Freeman of the City of London and proprietor of the notorious John Bull newspaper he was imprisoned three times for libel, declared a bankrupt, [and] became the manufacturer of the printing ink used to print the Penny Black stamp". In spite of Shackell's colourful background, it would seem that all of us who love nature and wildlife owe him a great debt of gratitude for securing and thereby protecting one of the last stretches of the River Lea to have mainly escaped the picks and shovels of the civil engineers.

When the anglers' club was first set up, the membership was made up of wealthy Victorian gentlemen who were mainly interested in mixed course fishing (particularly pike) as a sport. As the club developed, trout became the preferred fish and in the early twentieth century a fly-only rule for trout was enforced. Membership today has changed from the wealthy Victorian gentlemen founders to people from all walks of life and both sexes are included. To be a member today, all that is required, besides the membership fee, is a genuine love of fly fishing and a passion to conserve local fish stocks. The club has a rule that members can only take four trout during a week's fishing and further catches must be returned to the water to preserve fish stocks. Also, the club imposes a limit on membership numbers to around sixty so that over-fishing does not take place.

In recent years, concerns were raised over the dwindling amount of water passing through the River Lea loop as this was having a detrimental effect on the fishery. Water flow was down to a paltry 20 million litres per day. Diligent research into ancient documents by Feargal Sharkey discovered that over the years a number of agencies responsible for the maintenance of the Lee Navigation and other local waterways had been illegally starving the fishery

Henry Wix of Clay Hill House, Walthamstow, Honorary Secretary of Amwell Magna Fishery 1855–75.

of its rightful supply. When the matter was brought to the agencies' attention, work was put in hand to remedy the situation and the fishery now receives around 120 million litres of water per day, an increase of 500 percent.

Brown trout require very specific conditions for spawning and need to stay in a particular part of a river. Club members naturally share the trout's ambitions and have ensured that their two-mile stretch of river has all the facilities that the fish require to thrive. Over the years, thousands of tons of gravel have been obtained from nearby workings, before these sites became nature reserves, and this material was distributed along the river bed by club members to create riffles. These are small gravel bumps under the water which the trout can excavate to lay their eggs. To ensure that the water remains at the correct depth for the trout's liking, the club managed to secure several tons of granite blocks that were surplus to requirements after the demolition of John Rennie's London Bridge, built between 1824 and 1831. After demolition in the 1970s, the majority of the structure was purchased by an American businessman and shipped across the Atlantic Ocean to be rebuilt in the Arizona desert. The blocks, including a keystone, from one of the five arches of the bridge can now be seen a little below the waterline on a fast-flowing section of the river.

The river agencies have now recognised that the club is just not about the sport and pleasure of fishing but also a serious protector of water quality and a guardian of local flora and fauna. Recently, in recognition for this continuing stewardship, the club was provided with a substantial grant that will help them to hire specialist contractors with the necessary plant and equipment to repair and maintain eroding river banks.

When visiting the club in February 2017, I was shown around by the knowledgeable manager of the fishery, Bob Dear. As we walked along the river bank, I enquired about the function of the little covered platforms that

Part of the two-mile stretch of the River Lea below Ware that forms the Amwell Magna Fishery.

Ripples made on the surface as trout rise to a handful of fish bait.

Fly fishing for trout at Amwell Magna Fishery on the River Lea.

A fisherman showing his catch. This is a good indication of how clean the river is at Amwell Magna Fishery.

Bob Dear pointing to large blocks of granite below the waterline that came from the old London Bridge.

could be seen dotted along the river and raised above water level. Bob explained that these were mink platforms which had been covered with a material to record the presence of this voracious predator. Once the animal's footprints have been recorded, the material is smoothed over; if they occur again, a trap is placed on the platform. I understand from the LVRPA that minks have created a real problem for wildlife in the region.

Another voracious predator is the American signal crayfish which poses a massive threat to wildlife in the Lea Valley but also across Britain. This armour-plated monster was introduced into Britain in the 1970s as a food source, native crayfish stocks having been hit badly by a crayfish plague. What was not realised at the time was that the American signal crayfish was also a carrier of this plague. Unfortunately, Sweden and Finland had also introduced the predator and now the animal has spread right across Europe. Once these crayfish get into our waterways they consume almost everything in sight,

Bob Dear, manager of the Amwell Magna Fishery on a regular visit to inspect the site.

Below: *Bob Dear with an American Signal Crayfish trap. These devices are deployed to remove several tons of the invasive species each year.*

Below right: *Mink traps at Amwell Magna Fishery.*

taking fish eggs, small fish, insects and other small water creatures that are necessary to the native wildlife food cycle. The American signal crayfish burrow into river banks and cause massive erosion problems; and as if this weren't bad enough, the animal can travel across open ground to populate other water habitats. Last year alone, the Amwell Magna Fishery trapped over one ton of these wildlife-destroying monsters.

Besides these two major problems, mink and the American signal crayfish,

that the fishery has constantly to monitor, there is also another predator that walks on two legs. In recent years, Amwell Magna Fishery has been invaded not by introduced predators but by indigenous poachers. These uncaring people have absolutely no thought for wildlife conservation and protection. Apart from littering the fishery with empty beer bottles and other detritus, they also took trout that were ready to spawn and even removed American signal crayfish from traps. Now the fishery has had to invest in high-definition secret camera traps to catch any future culprits. It is a sad reflection on our society that there are people, only interested in their own selfish gratification, who are prepared thoughtlessly to undo years of conservation work by those who care for our planet.

However, the good news is that Amwell Magna Fishery is currently run by a group of enthusiastic members and this gives me great confidence that this little-known jewel in the Lea Valley's crown has a bright future.

REFERENCES

Dear, Bob, personal conversation during visit to Amwell Magna Fishery, February 2017

Sharkey, Feargal, *A Tale of Two Weirs: How One of the 18th Century's Leading Civil Engineers Put a River in the Wrong Place*, Amwell Magna Fishery, 2016

Sharkey, Feargal, personal conversation, January 2017

THE STORY OF A LEA VALLEY VIKING BOAT

Today, from across the Atlantic, we hear from a certain person, whose name will not be mentioned in this book, complaints against the media of "fake news". However, such charges are not particularly new phenomena. As deadlines beckon, journalists and broadcasters can sometimes exaggerate or fail to diligently check their facts or sources. On the whole, our respected media sources in the UK provide a useful information service, allowing us to question, interpret and, importantly, read between the lines. The story that is about to unfold, which began in the early twentieth century, possesses many of the elements just discussed.

The Walthamstow boat excavated from the Lockwood Reservoir in June 1900.

The Lockwood and Banbury reservoirs were constructed for the East London Waterworks Company (ELWC) between 1899 and 1903 and are situated north of Ferry Lane, Walthamstow and east of Tottenham Marsh. On 11th June 1900, a Mr C.W. Sharrock, Clerk of Works for S. Pearson & Son Ltd which was the on-site contractor carrying out the work, wrote a letter to Mr W.B. Bryan, the engineer for the ELWC, to report the following: "In excavating near the north east corner of 'Lockwood' Reservoir, we have unearthed what appears to be an old Viking ship. From the formation of the ground above it, it must have been here for some hundreds of years. We have asked Mr Romilly Allen, who is an authority on these matters, to come down and see it, and we expect him here on Wednesday next."

The Walthamstow boat was said to measure 50ft in length and 26ft in width.

After this initial correspondence, letters begin to flow between the contractor and the ELWC, and various experts were invited to visit the site. These included the Secretary of the Essex Field Club and a representative from the British Museum. Presumably, these people wrote reports after their site visits which could be hidden away in some dusty archive. If so, there is hope that they might be discovered today by an interested researcher.

On 5th July 1900, the *Globe* newspaper appears to have been one of the first to go live with the story and published the following account:

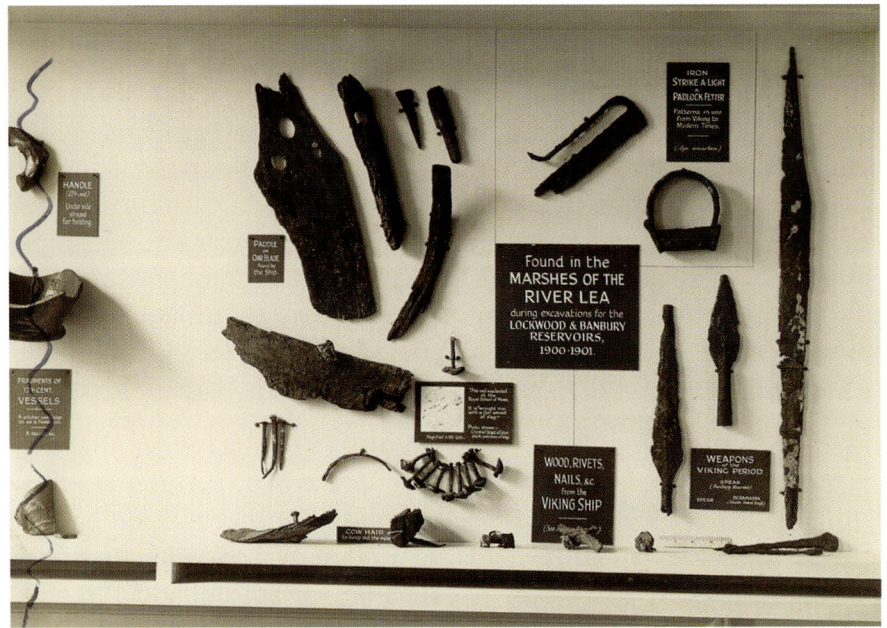

An aerial view of the Walthamstow reservoirs looking north with the Lockwood Reservoir centre picture at the top of the grouping.

A collection of artefacts collected from the Lockwood Reservoir at the time of the excavation in 1900.

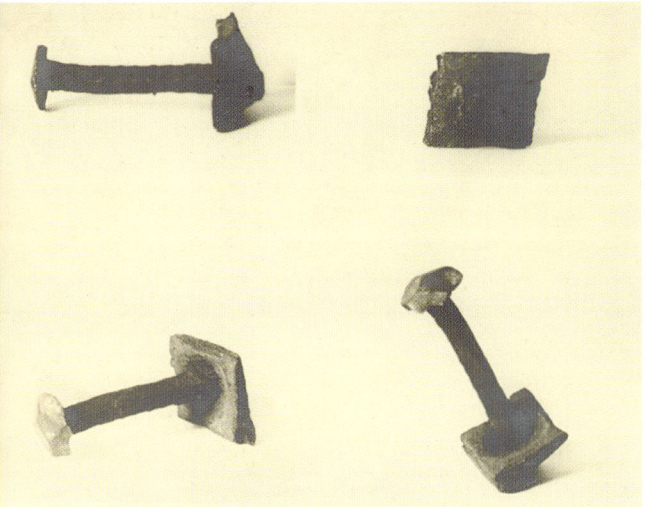

Rivets from the Walthamstow boat recovered during the excavated in 1900.

Pieces of the Walthamstow boat that are stored at Bruce Castle Museum.

The East London Water Company in excavating their new reservoirs at Tottenham Marshes have made a discovery of unique interest. A war vessel 50 feet long with a beam of 26 feet, made of oak and elm, has been dug up in an almost complete state. From several special indications archaeologists claim to give the exact date of the vessel. The form of the rivets prove that she is of Danish build, and it is not an outrageous inference to argue that the ship belonged to the Danes who were defeated by King Alfred in the Lea Valley in AD 894. At any rate the conjecture is plausible and a somewhat rare occurrence. Archaeologists are in agreement.

Following this newspaper report, numerous stories appeared in local and national newspapers; letters also abounded between the ELWC, contractors, archaeologists, museum experts and interested members of the public. This led to a confusing mix of correspondence and reports, provoking articles about the find to be written in the 1920s, 1930s and 1960s by later researchers who were trying to unravel the truth behind the 'Walthamstow boat' story. Some of the stories that emerged after the *Globe* report have suggested that the newspaper article had motivated a large crowd to invade the reservoir site and the police were rendered powerless to stop the invaders carrying off fragments of the wreck as souvenirs. After this event, correspondence from the ELWC suggests that the Company was so appalled by what had happened that they promptly threw a cloak of secrecy over the boat and also placed an embargo upon their workmen to remain silent on past and new discoveries.

As a result of this action, some later researchers have complained that they have been denied the opportunity to examine crucial evidence, especially for one story which appears to have come from the contractor's workmen. This has it that the boat was found keel up, covering a skeleton, gold ornaments and an iron Viking sword, which would have probably indicated a Viking boat burial. The story then takes another incredible twist. A letter from the famous archaeologist, the late Sir Mortimer Wheeler to a G. E. Roebuck, dated 3rd March 1932, states that:

> The Viking sword for which you are searching is now in the collection of Prince Ladislaus Odescalchi in Rome. It is illustrated on page 16 of Volume I of Laking's 'Arms and Armour', and also on page 34 of the London Museum guide 'London and the Vikings'. The reference to 'Laking' is to Sir Guy Laking who somehow acquired the sword for his personal collection, then at some time, he allegedly passed it on to Prince Ladislaus Odescalchi for his collection in Rome.

By now, you may be wondering why I have introduced this story into a book that is primarily about learning the lessons of how industrial development destroyed natural wildlife habitats, causing major problems for our planet; and how we are now learning to put these delicate and complicated problems right.

First, I wanted to use this publication to put the Walthamstow boat story back on the public agenda in the twenty-first century because there is a much bigger story that emanates from the Lea Valley region that has never been properly researched. Now we have the opportunity to use the wealth of modern technological tools that are available to archaeologists, historians and researchers to attempt to solve a story that is a mere 1,123 years old.

Since leaving industry, I have mainly pursued a career of researching, writing, teaching and broadcasting about my specialist subject: the industrial history of London's Lea Valley. However, while researching within the region, stories often emerge that are not strictly within my subject area. I have always been puzzled by the fact that local archaeological societies, historical societies and museums are not able to provide convincing evidence that a major battle took place in the area when a Viking fleet sailed up the River Lea and invaded English territory (mentioned earlier in the chapter 'Time Travelling through London's Lea Valley from Pre-history to the Present').

According to the Anglo Saxon Chronicle, in the year 894 AD a large Viking fleet sailed up the River Thames and then up the River Lea to a point twenty miles north of London (this would place the fleet around the town of Ware, Hertfordshire) where a fortified camp was built with ditches and ramparts. It is recorded that King Alfred (of Wessex) deprived the Danes of escape by blocking, or possibly lowering, the river. The Danes abandoned their camp, leaving their boats behind, and escaped by sending their women and children across country to East Anglia while the men marched overland to Bridgenorth in Shropshire.

As the River Lea formed part of the boundary between Danelaw and Wessex, we now have a connection with the Walthamstow boat. If an examination of the remaining timbers and iron rivets (which are located in local and national museums) could take place, it should be possible to determine, one way or another, whether the boat could have been part of the invasion fleet. In this way, the whole story becomes an important chapter in the making of Britain. This could also provide the opportunity to bring the story to the attention of the public if a broadcaster could be persuaded to make a television documentary by filming the investigation and research programme as it unfolded. Furthermore, the publicity gained for the Lea Valley from a full-length TV documentary would help make up for the many years that the region has been overlooked as an important part of Britain's social and industrial heritage, and the birthplace of the post-industrial technological revolution.

In highlighting the Viking story, I hope to encourage schools, colleges, universities, researchers, historians, archaeologists and television companies to come together as a structured team to carry out the necessary preparations to identify appropriate archaeological sites that should be investigated. This would give television producers and camera crews the opportunity to film work as it progressed and build up material for the TV documentary. Hopefully, someone with influence will read this section of the book. If this should happen, it would be a wonderful legacy!

The second point is that the story of the Walthamstow boat is a good example of how history becomes muddled and distorted if we do not have guidelines in place that allow professionals to be called in to meticulously record and conserve artefacts that are important to our heritage. Of course, we do have this luxury today. When a new building development takes place, although there have been occasions when some uncaring individuals have tried to avoid complying, there is a legal requirement for the site to be examined by professional archaeologists before building work can commence.

Now, for a moment, let us consider the parallels between the Walthamstow boat story and the present desire, as highlighted throughout this book, to protect our planet by trying to reverse, even at this late stage, the damage we have caused to natural habitats through our desire for more industrialised goods. We now have the science and understanding that our forefathers lacked and we have finally learned how to reduce CO_2 pollution. Already we have systems in place that can produce clean energy, create safe methods of animal husbandry, agriculture and horticulture and we also have the ability to limit the use of pesticides and insecticides which can poison humans and wildlife when they enter our bodies through skin, air and the food chain. We should be looking for natural solutions, some of which are currently available, to overcome these problems. Also, governments across the world should be funding and encouraging ongoing scientific research to find new ways of reducing the use of all harmful chemicals, effluents and gasses that affect the delicately balanced ecosystems of our rapidly decaying planet.

REFERENCES

Authors unknown, *Anglo-Saxon Chronicle*, Macmillan, 1982, translated by Ann Savage

Bruce Castle Museum, Tottenham, archive collection of letters, articles and documents on the Walthamstow Boat

Clark, John, Former curator of the Museum of London and organiser of the King Alfred Exhibition, personal correspondence, February 2017

Heals, Gary, Curator, Vestry House Museum, private conversation, November 2016

Hedgecock, Deborah, Curator, Bruce Castle Museum, private conversation, February 2017

Vestry House Museum, Walthamstow, archive collection of letters, articles and documents on the Walthamstow Boat

THE QUEEN ELIZABETH OLYMPIC PARK

On 6th July 2005 in Singapore, the International Olympic Committee announced that London had won the bid to host the Olympic and Paralympic Games in the year 2012. As the result was announced, television crews began interviewing weary, yet ecstatic British officials, who up until then had spent every available moment putting the finishing touches to their bid presentation. Not all, it would seem, had believed that London could pull off this nail-biting win against what many pundits claimed to be the favoured city, Paris. When the television cameras cut to the waiting crowds at Stratford, in London's East End, watchers could not fail to be moved by the scenes of sheer excitement as local residents came to realise that the Games' major facilities would be located on their doorstep in the lower Lea Valley, an area that over the years had seen more than its fair share of industrial dereliction and neglect. For decades, the region had been crying out for a regeneration stimulus of this magnitude and it would seem that the dreams of local people were about to be realised.

Looking south across the Olympic Park with the River Lea centre picture.

Children enjoying a play and explore area on the Olympic Park.

The construction of the Queen Elizabeth Olympic Park to accommodate the 2012 Games turned out to be a magnificent opportunity to regenerate not only Stratford, but also the surrounding East End community. This area of east London had remained, for many years, neglected and unloved. As Britain's industrial base, along with the Thames Docklands region, had been in serious decline since the 1970s, there was not a faint glimmer of hope that things would change economically for the better in the foreseeable future. It is probably fair to say that the region had never fully recovered from the effects of the Blitz and the unimaginative architectural developments such as the high-rise blocks of flats that were erected in the 1960s. These high-rise buildings had been constructed with little thought for the well-being of their residents and had isolated members of once tight-knit communities that had grown together and supported each other during the war years.

Artist's impression of the London Olympics Aquatics Centre.

Artist's impression of the Olympic village.

Perhaps the vision for the current, and ongoing, transformation of Stratford and its East End neighbours can best be summed up by the following statement from the London Legacy Development Corporation: "the ultimate goal is that the Queen Elizabeth Olympic Park becomes a natural extension of its environment, and the boundaries between the Park and its neighbouring communities no longer exist."

Once decontamination of the highly polluted Olympic site and the removal of the electricity pylons had taken place, architects, planners and developers then had the momentous task of creating a twenty-first-century landscape around a tangle of railway lines, waterways and Bazalgette's Victorian sewage system, all of which criss-crossed the Park like a badly knitted pullover made from spaghetti. Those who had grown up and lived in the Stratford area for most of their lives could not have failed to have been overwhelmed and astounded by the complete transformation of a once neglected and derelict landscape, that had previously supported scrap-yards, old and smelly industries like bone renders, varnish works, oil refiners, paint manufacturers and sulphuric acid distillers, in such a relatively short time. In the words of the authors of *The Making of the Queen Elizabeth Olympic Park*, the project came in "on time, within budget, to multiple stakeholders' satisfaction, and to challenging targets for sustainability, [and] the Olympic Delivery Authority (ODA) established a new benchmark for development everywhere".

Although the redevelopment of the Park presented the planners with a massive headache, it also provided a unique opportunity to apply modern sustainable eco-friendly solutions to a major city landscape development on a scale that had probably never been attempted before. As pointed out earlier in this book, since the construction of Bazalgette's sewage system in the mid-nineteenth century and the subsequent growth of London, land within the River Lea and Thames flood plains has been built on and concreted over. During heavy downpours surface water is unable to escape so drains and sewers back-up and effluent discharges have to be made into the local water courses that are connected to the Thames. For a country that was about to accommodate and welcome the world's top athletes to London, especially with all the media attention that the Games would bring, this was not the pre-Olympic publicity that Britain needed.

It was planned that the 250 hectare Park would accommodate the Olympic Stadium and a separate warm-up track, Velodrome, Copper Box arena, Aquatics Centre, BMX Track, Eton Manor area for Paralympic tennis, Riverbank Arena with sponsors' hospitality facility, the Orbit Tower, Athletes' Village, Press Centre and the International Broadcast Centre. Other amenities were also created to accommodate paved and landscaped public access, operational facilities and transport malls.

Therefore, the planners of the Park did not need to be reminded that their first priority was to design a scheme for the site that would cope with a worst downpour scenario. After all, they were well aware that the Thames Water

Olympic stadium with wildflower meadows in foreground.

four-mile spur tunnel (Lee Tunnel) from Abbey Mills Pumping Station, Stratford to Beckton Sewage Treatment Works, and the Thames Tideway Tunnel project, were only at the planning stage and were not due to be completed until some years after the 2012 Games were over. Clearly, it would appear, choices to protect the Park from flooding were decidedly limited.

Olympic Park wildflower gardens.

It was known that the fluvial[16] peak within the Lower Lea Valley typically occurred around twenty-four hours after a heavy downpour and so a solution was finally approved, between the planners, the Environment Agency and British Waterways, that surface water runoff from the Park could be collected and discharged into the site's dredged and improved watercourses ahead of the river's peak flow. This action was combined, during the Park's development phase, with changes to the site's topography that raised certain areas around nine metres and by the forming of plateaus for the various venues that put them above river flood levels. Also, a reed-bed wetland area was created by widening part of the River Lea that flowed through the Park.

One of the Olympic Park's play and explore areas.

The creation of the Park was not just for the period of the Olympics. A long-term plan had been devised by the relevant agencies to ensure that a mixed-use development was created with extensive landscaping, where possible including green-roofed buildings to encourage bio-diversity. It was also planned to retain several Olympic venues which, after conversion, were designated to become sustainable legacy projects. To achieve these long-term goals, considerable forethought had gone in to the early planning stage to try and ensure that major unforeseen overspends would not occur. However, apart from all the forward thinking, it was reported that senior voices from within the athletics fraternity had insisted that the Olympic Stadium remained only for the staging of athletic events after the Games which removed the need to build in stadium flexibility during the design stage. This, in my view, was a grave mistake: we have now witnessed the conversion of the stadium, at great expense, to accommodate West Ham United Football Club and to make the stadium suitable for other commercial activities.

The Olympic Stadium with the Orbit in the foreground.

Nevertheless, the early planners had got their post-Olympic schemes spot on by ensuring pedestrian concourse areas of the Park were laid with porous asphalt strips that would allow water

Rainwater harvesting tanks below ground.

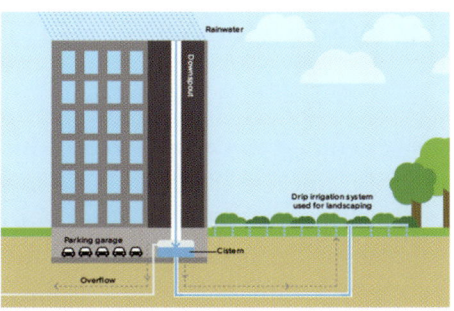

Rainwater collection diagram from high-rise buildings.

Old Ford water recycling plant.

runoff into below-ground granular trenches containing perforated pipes. In turn these pipes would convey the water to catch-pits where discharge could take place into adjacent water courses. Also a network of underground irrigation pipes were laid to feed the Park's extensive landscaped gardens and flower meadow quarter, and in the wetland areas small volume balancing ponds were constructed to complement the water management system.

Another important part of the of the Olympic Park's water strategy is the Old Ford Water Recycling Plant. Located towards the northern end of the Park in a secluded wooded area with mature trees, a pond and wild grasses, the larch-clad building houses an innovative treatment plant that processes 600,000 litres of reclaimed water per day. Raw sewage is taken from Bazalgette's Northern Outfall Sewer and fed into two underground septic tanks for primary settling before the effluent passes through multi-layered filters and is treated by a membrane bioreactor (MBR) process. Further processing is required to remove residual colour and chlorine dosing ensures suppression of bacterial re-growth. The final act is to feed the reclaimed water to an external larch-clad storage tank from where it can be pumped to various facilities around the Park to flush toilets. After the Games, some venues will retain this water-flushing facility while Thames Water is committed to running the plant for seven years so that data can be gathered to establish if it is economical to utilise this method of water recycling for other purposes.

REFERENCES

Author unknown, Olympic Park London,
   http://susdrain.org/case_studies/olympic_park_london.html
Birch, Amanda, Olympic Treatment, *Architects' Journal*, 26th July 2012
Brickell, Paul, Director of Regeneration, Olympic Legacy Development
   Corporation, personal conversation, March 2017
Hopkins, John, and Neal, Peter, *The Making of the Queen Elizabeth Olympic
   Park*, Wiley & Sons Ltd, Publication 2013
Lewis, Jim, *Water and Waste: Four Hundred Years of Health Improvements in
   the Lea Valley*, Middlesex University Press, 2009
Scott, Troy, Personal Assistant, Olympic Legacy Development Corporation,
   personal conversation, March 2017

THE WALTHAMSTOW WETLANDS PROJECT

The Walthamstow Wetlands project is scheduled to be completed this year (2017), on a site of Special Scientific Interest (SSSI) in London's Lea Valley. When this happens, the British public will be gifted with one of the largest urban wetland nature reserves in Europe. Funding for the project has come from a number of sources, £4.47m from the Heritage Lottery Fund, £0.75m from the Greater London Authority and £1.84m from Thames Water, the site owners. Further good news is that the Greater London Authority have agreed to fund the development of a wetlands-to-wetlands greenway by improving the three-kilometre cycle link between Woodbury Wetlands (near Manor House, Haringey) and the Walthamstow Wetlands site. It is hoped that this cycleway will encourage visitors to explore both sites, which could easily be achieved in a single day by non-polluting transport.

The Marine Engine House, formerly called the Ferry Lane Pumping Station, got its name when the original steam engine was replaced by a powerful and more efficient ship's engine. This locally listed pumping station is being transformed into a visitor centre with exhibition space, a café and a viewing terrace. In the long term it is planned to rebuild the pumping station's chimney, which was demolished after the marine engine was installed; this will be done to create homes for visiting swifts.

Educational facilities for schools and the community will be provided that will enhance our knowledge and encourage us to become protectors and conservers of nature rather than uncaring bystanders. This will be an extremely important facility as there are still, within our society, the hangovers from the age of the great Victorian collectors who decimated much of the world's wildlife so that they could compete against each other in the race for the rarest or most magnificent collections. Nothing seemed to have escaped their

Contractor's boat during the development of the Walthamstow Wetlands.

Artist's impression of how the Walthamstow Wetlands project will look when completed.

Early reed-bed planting at the Walthamstow Wetlands.

Copper tokens that were once produced at Coppermill. The building is now part of the Walthamstow Wetlands complex.

attention, as can be witnessed in the many natural history museum collections around the world, with great showcases literally stuffed with fish, reptiles, animals, birds and butterflies. Even today we have trophy hunters who kill endangered species for sport. And of course there are those that kill and trap animals like rhino and tiger for their body parts to supply the network of criminals that operate the supply chain which particularly feeds the spurious Oriental and African medicinal trade. Then there is the slaughter of elephants for ivory and also the capture of birds, reptiles and primates for domestic pets. All these practices must be debated within our communities and then internationally outlawed as future generations must understand the delicate and intricate working of our ecosystems.

Many readers will recall the years after the Second World War when it was still thought normal for children to collect and swap birds' eggs, catch butterflies and moths for their display cabinets and spend endless days during school holidays netting frog spawn, tadpoles and minnows then taking them home in jam jars. Usually the contents had to be tipped away when the child got home as there was no place where these specimens could be kept in a suitable environment. All this was done in the name of enjoyment and fun. Parents did not explain to their children that these activities were wrong as they themselves had participated in such deeds in their childhood, as had their parents before them.

It is planned that the beautiful Italianate Coppermill pumping station, located to the south east of the Wetlands project site and currently used as a storeroom for Thames Water equipment, will be adapted as a viewing platform allowing visitors spectacular views across the wetlands and the marshes. There will be free public entrance to the site during opening hours and it is planned to have four new entrances that will give greater access to foot and cycle pathways.

Even though the Walthamstow Wetlands project has yet to be completed, the site is already home to a whole range of breeding birds and wintering wildfowl including Gadwall, Pochards and Shovelers. The site also has breeding populations of Cormorants, Grey Herons, Little Egrets, Tufted Duck, Coot, Moorhen and other waterfowl. Wading birds like Little Stint, Golden Plovers, Whimbrel, Wood Sandpipers, Oystercatcher, Snipe, Ruff, Curlews, Ringed and Little Ringed Plovers, Lapwings, Redshanks, Dunlin, Green and Common Sandpipers are attracted to the site and will often take the opportunity to stop and take a break to feed and refresh on their long-distance journeys. In the long term, it is hoped that the extra facilities being created

Egyptian Goose on the Walthamstow Wetlands.

Redshanks at Walthamstow Wetlands.

One of two cormorant islands at Walthamstow Wetlands.

Tufted Duck at Walthamstow Wetlands.

Geese on a calm day at Walthamstow Wetlands.

for wildlife at the Walthamstow Wetlands will encourage some of these bird species to over-winter here. It might also be possible to encourage some of the offspring of the elusive northern Lea Valley Bitterns to take up residence in the newly created reed beds on the Ferry Lane site.

The Walthamstow Wetlands project is being delivered by the London Wildlife Trust, which will manage the day-to-day running of the site when it is completed, in partnership with the London Borough of Waltham Forest and Thames Water. Let us hope that this visionary project lives up to our high expectations by receiving multiple visits from schools, colleges, universities, community groups and members of the public so that the message of conserving our fragile ecosystem becomes part of everyone's DNA. Future generations must learn never to turn their backs on our natural world ever again.

REFERENCES

Author unknown, *Lea Bridge Road Planning Framework*, Urban Practitioners, October 2009

Author unknown, *Upper Lea Valley Landscape Strategy*, Witherford Watson Mann Architects, February 2010

Author unknown, *Walthamstow Reservoirs Feasibility Study*, Chris Blandford and Associates, April 2010

Author unknown, *Draft Upper Lea Valley Planning Framework*, Greater London Authority, January 2011

CONCLUSION

When thinking about a conclusion for this book, a front-page article in the *Guardian* – 'Toxic air risk to one in four London schools' – caught my eye. In the article it was alleged that a report had been kept secret by the former Mayor of London which revealed in 2016 that 433 primary schools were exposed to dangerous levels of air pollution. The article went on to explain that new data shows "802 out of 3,261 nurseries, primary and secondary schools and higher education colleges are within 150 metres of NO_2 pollution levels that exceed the EU legal limit of 40 micrograms per cubic metre of air". In an interview, Dr Francis Gilchrist, consultant respiratory paediatrician at Royal Stoke University hospital, said "it was known that children were particularly sensitive to air pollution and that lung damage had lifelong consequences".

It is known that traffic in the capital is a major contributor to air pollution but only recently has the focus turned to diesel vehicles, which contribute through the production of particulate matter and nitrogen oxides (NO_X). It seems extraordinary that we have only in recent years been able to link increasing levels of respiratory illnesses like chest infections and asthma in children with the air pollution that is exacerbated by diesel usage. It was not long ago that vehicle manufacturers and government agencies were recommending that the public buy diesel cars, vans and lorries on the grounds that they were more efficient than their petroleum counterparts. Now the recommendations are to take diesel vehicles off our roads altogether!

These problems of air pollution have made me think about their impact upon nature and the other conservation issues that have been discussed throughout this book. Therefore, I decided to see if scientists around the world had produced any evidence that these man-made pollutants were having, or not having, a good or a bad effect upon plants and wildlife. A quick search on the Internet soon found a plethora of papers, articles and reports which collectively showed that the answers to my question were not completely straightforward. However, it is generally acknowledged that different chemical contaminants within airborne pollutants can cause growth problems and also mutations in different types of plant and wildlife that can each react in different ways. For example, when a pollutant like sulphur dioxide is released into the atmosphere, and then combines with water droplets in the clouds, it can return to earth as acid rain. When acid rain falls over a large forested area, it can kill trees or stunt their growth. On infiltrating soil, acid rain will change the chemistry and make the ground unfit for plants, insects and animals that have made the soil a habitat. The contaminated water runoff gets into rivers, lakes and streams causing untold damage to plants, fish and other aquatic wildlife. Of course, rivers run to the sea, so the contamination then becomes a global problem affecting marine life.

Many cities across the developed world have put in place programmes to monitor levels of air pollution and have found that large amounts of fine particulate matter have been shown to shorten life spans by causing cardiovascular and pulmonary problems. One component of this fine particulate matter is black carbon, which is emitted particularly from diesel vehicles. It is claimed that black carbon is responsible for 2.1 million premature deaths annually.

Scientists are producing more and more evidence showing how man-made contaminates are damaging our planet, our plants, our wildlife and ourselves. However, all the mechanisms for this are not yet fully understood. This has given the deniers the opportunity to voice their opinions that global warming and associated changes to weather patterns are probably a figment of our scientists' imaginations. My response to this would be: what is the harm in trying to reduce CO_2 and other harmful emissions to pre-Industrial Revolution levels?

REFERENCES

Ashden, T.W., Ashmore, M., Bell, J.N.B., Bignal, K., Binnie, J., Cape, J.N., Caporn, S.J.M., Carroll, J., Davison, A., Hadfield, P., Honour, S., Lawton, K., Moore, S., Power, S., and Shields, C., Centre for Ecology and Hydrology, UK, University of Bradford, UK, Imperial College, UK, Manchester Metropolitan University, UK, Newcastle University, UK, 'Impact of vehicle emissions on vegetation', *Transactions of the Built Environment*, Vol. 64, 2003

'Toxic air risk to one in four London schools', *Guardian*, 25th February 2017

Williams, Florence, 'Warning: Living in a City Could Seriously Damage Your Health', *Guardian*, 13th March 2017

APPENDIX — THE WILDLIFE TRUSTS

Founded in 1912 by Charles Rothschild, the English banker and entomologist, and originally called the Society for the Promotion of Nature Reserves (SPNR), the trading name of this organisation is now the Royal Society of Wildlife Trusts (RSWT). Familiarly referred to as the Wildlife Trusts, the organisation is currently made up of 47 local Wildlife Trusts from across the UK, the Isle of Man and Alderney in the Channel Islands. Between them, the Wildlife Trusts are responsible for overseeing some 2,300 nature reserves that cover over 90,000 hectares. The Trusts currently have around 80,000 members. Charles, Prince of Wales, a royal with nature and wildlife in his heart, is patron of the Wildlife Trusts and the current president is the highly respected naturalist, writer and campaigner Tony Juniper who was appointed in December 2015. Vice presidents are Sir David Attenborough, Chris Baines, David MacDonald, Julian Pettifer, Robert Worcester, Chris Packham, Nick Baker, Bill Bolsover and Bill Oddie.

REFERENCES

Author unknown, *Biodiversity Benchmark*, Wildlife Trusts, April 2012

Author unknown, *The Rothschild Reserves*, Wildlife Trusts, December 2014